Handbook of Fiber Science and Technology: Volume I

Chemical Processing of Fibers and Fabrics

FUNDAMENTALS AND PREPARATION

Part A

INTERNATIONAL FIBER SCIENCE AND TECHNOLOGY SERIES

Handbook of Fiber Science and Technology: Volume I

Chemical Processing of Fibers and Fabrics
FUNDAMENTALS AND PREPARATION Part A

edited by Menachem Lewin and Stephen B. Sello

Chemical Processing of Fibers and Fabrics
FUNDAMENTALS AND PREPARATION Part B (in press)

edited by Menachem Lewin and Stephen B. Sello

Handbook of Fiber Science and Technology: Volume II

Chemical Processing of Fibers and Fabrics
FUNCTIONAL FINISHES Part A

edited by Menachem Lewin and Stephen B. Sello

Chemical Processing of Fibers and Fabrics
FUNCTIONAL FINISHES Part B (in press)

edited by Menachem Lewin and Stephen B. Sello

Other volumes in preparation

Handbook of Fiber Science and Technology: Volume I

Chemical Processing of Fibers and Fabrics

FUNDAMENTALS AND PREPARATION

Part A

edited by

Menachem Lewin
Israel Fiber Institute
and Hebrew University
Jerusalem, Israel

Stephen B. Sello
J. P. Stevens & Co., Inc.
New York, New York
and Greenville, South Carolina

MARCEL DEKKER, INC. New York and Basel

Library of Congress Cataloging in Publication Data
Main entry under title:

Handbook of fiber science and technology.

(International fiber science and technology series)
Includes bibliographical references and indexes.
Contents: v. 1. Chemical processing of fibers and fabrics—fundamentals and preparation
1. Textile finishing. I. Lewin, Menachem, [date]. II. Sello, Stephen B., [date]. III. Series.
TS1510.H3 1983 677'.02825 83-7685
ISBN 0-8247-7010-2 (v. 1, pt. A)

MARCEL DEKKER, INC.
270 Madison Avenue, New York, New York 10016

Current printing (last digit):
10 9 8 7 6 5 4 3 2 1

PRINTED IN THE UNITED STATES OF AMERICA

ABOUT THE SERIES

When human life began on this earth *food* and *shelter* were the two most important necessities. Immediately thereafter, however, came *clothing*. The first materials used for it were fur, hide, skin, and leaves—all of them sheetlike, two-dimensional structures not too abundantly available and somewhat awkward to handle. It was then—quite a few thousand years ago—that a very important invention was made: to *manufacture* two-dimensional systems—fabrics—from simple monodimensional elements—fibers; it was the birth of textile industry based on fiber science and technology. Fibers were readily available everywhere; they came from animals (wool, hair, and silk) or from plants (cotton, flax, hemp, and reeds). Even though their chemical composition and mechanical properties were very different, yarns were made of the fibers by spinning and fabrics were produced from the yarns by weaving and knitting. An elaborate, widespread, and highly sophisticated art developed in the course of many centuries at locations all over the globe virtually independent from each other. The fibers had to be gained from their natural sources, purified and extracted, drawn out into yarns of uniform diameter and texture, and converted into textile goods of many kinds. It was all done by hand using rather simple and self-made equipment and it was all based on empirical craftsmanship using only the most necessary quantitative measurements. It was also performed with no knowledge of the chemical composition, let alone the molecular structure of the individual fibers. Yet by ingenuity, taste, and patience, myriads of products of breathtaking beauty, remarkable utility, and surprising durability were obtained in many cases. *This first era* started at the very beginning of civilization and extended into the twentieth century when steam-driven machinery invaded the mechanical operations and some empirical procedures—mercerization of cotton, moth-proofing of wool and loading of silk—started to introduce some chemistry into the processing.

The second phase in the utilization of materials for the preparation and production of fibers and textiles was ushered in by an accidental discovery which Christian Friedrich Schoenbein, chemistry professor at the University of Basel in Switzerland, made in 1846. He observed that cotton may be converted into a soluble and plastic substance by the action of a mixture of nitric and sulfuric acid; this substance or its solution was extruded into fine filaments by Hilaire de Chardonnet in 1884.

Organic chemistry, which was a highly developed scientific discipline by that time, gave the correct interpretation of this phenomenon: the action of the acids on cellulose—a natural fiber former—converted it into a *derivative*, in this case into a cellulose nitrate, which was soluble and, therefore, spinnable. The intriguing possibility of manipulating natural products (cellulose, proteins, chitin, and others) by chemical action and thereby rendering them soluble, resulted in additional efforts which led to the discovery and preparation of several cellulose esters, notably the cellulose xanthate and cellulose acetate. Early in the twentieth century each compound became the basis of a large industry: viscose rayon and acetate rayon. In each case special processes had to be designed for the conversion of these two compounds into a fiber, but once this was done, the entire mechanical technology of yarn and fabric production which had been developed for the natural fibers was available for the use of the new ones. In this manner new textile goods of remarkable quality were produced, ranging from very shear and beautiful dresses to tough and durable tire cords and transport belts. Fundamentally these materials were not truly "synthetic" because a known natural fiber former—cellulose or protein—was used as a base; the new products were "artificial" or "man-made." In the 1920s, when viscose and acetate rayon became important commercial items polymer science had started to emerge from its infancy and now provided the chance to make *new fiber formers* directly by the polymerization of the respective monomers. Fibers made out of these polymers would therefore be "truly synthetic" and represent additional, extremely numerous ways to arrive at new textile goods. Now started the *third era* of fiber science and technology. First the basic characteristics of a good synthetic fiber former had to be established. They were: ready spinnability from melt or solution; resistence against standard organic solvents, acids, and bases; high softening range (preferably above 220°C); and the capacity to be drawn into molecularly oriented fine filaments of high strength and great resilience. There exist literally many hundreds of polymers or copolymers which, to a certain extent, fulfill the above requirements. The first commercially successful class were the *polyamides*, simultaneously developed in the United States by W. H. Carothers of duPont and by Paul Schlack of I. G. Farben in Germany. The *nylons*, as they are called commercially, are still a very important class of textile fibers covering a remarkably wide range of properties and uses. They were soon (in the 1940s) followed by the *polyesters*, *polyacrylics*, and *polyvinyls*, and somewhat later (in the 1950s) there were added the *polyolefins* and *polyurethanes*. Naturally, the existence of so many fiber formers of different chemical composition initiated successful research on the molecular and supermolecular structure of these systems and on the dependence of the ultimate technical properties on such structures.

As time went on (in the 1960s), a large body of sound knowledge on structure-property relationships was accumulated. It permitted embarkation on the reverse approach: "tell me what properties you want and I shall *tailor-make* you the fiber former." Many different techniques exist for the "tailor-making": graft and block copolymers, surface treatments, polyblends, two-component fiber spinning, and cross-section modification. The systematic use of this "macromolecular engineering" has led to a very large number of *specialty fibers* in each of the main classes; in some cases they have properties which none of the prior materials—natural and "man-made"—had, such as high elasticity, heat setting, and moisture repellency. An important result was that the new fibers were not content to fit into the existing textile machinery, but they suggested and introduced substantial modifications and innovations such as modern high-speed spinning, weaving and knitting, and several new technologies of texturing and crimping fibers and yarns.

This third phase of fiber science and engineering is presently far from being complete, but already a *fourth era* has begun to make its appearance, namely in fibers for uses *outside* the domain of the classical textile industry. Such new applications involve fibers for the reinforcement of thermoplastics and duroplastics to be used in the construction of spacecraft, airplanes, buses, trucks, cars, boats, and buildings; optical fibers for light telephony; and fibrous materials for a large array of applications in medicine and hygiene. This phase is still in its infancy but offers many opportunities to create entirely new polymeris systems adapted by their structure to the novel applications outside the textile fields.

This series on fiber science and technology intends to present, review, and summarize the present state in this vast area of human activities and give a balanced picture of it. The emphasis will have to be properly distributed on synthesis, characterization, structure, properties, and applications.

It is hoped that this series will serve the scientific and technical community by presenting a new source of organized information, by focusing attention to the various aspects of the fascinating field of fiber science and technology, and by facilitating interaction and mutual fertilization between this field and other disciplines, thus paving the way to new creative developments.

Herman F. Mark

INTRODUCTION TO THE HANDBOOK

The Handbook of Fiber Science and Technology is composed of five volumes: chemical processing of fibers and fabrics; fiber chemistry; specialty fibers; physics and mechanics of fibers and fiber assemblies; and fiber structure. It summarizes distinct parts of the body of knowledge in a vast field of human endeavor, and brings a coherent picture of developments, particularly in the last three decades.

It is mainly during these three decades that the development of polymer science took place and opened the way to the understanding of the fiber structure, which in turn enabled the creation of a variety of fibers from natural and artificial polymeric molecules. During this period far-reaching changes in chemical processing of fabrics and fibers were developed and new processes for fabric preparation as well as for functional finishing were invented, designed, and introduced. Light was thrown on the complex nature of fiber assemblies and their dependence on the original properties of the individual fibers. The better understanding of the behavior of these assemblies enabled spectacular developments in the field of nonwovens and felts. Lately, a new array of sophisticated specialty fibers, sometimes tailor-made to specific end-uses, has emerged and is ever-expanding into the area of high technology.

The handbook is necessarily limited to the above areas. It will not deal with conventional textile processing, such as spinning, weaving, knitting, and production of nonwovens. These fields of technology are vast, diversified, and highly innovative and deserve a specialized treatment. The same applies to dyeing, which will be treated in separate volumes. The handbook is designed to create an understanding of the fundamentals, principles, mechanisms, and processes involved in the field of fiber science and technology; its objective is not to provide all detailed procedures on the formation, processing, and modification of the various fibers and fabrics.

Menachem Lewin

INTRODUCTION TO VOLUMES I AND II

Textiles have undergone wet chemical processing since time immemorial. Human ingenuity and imagination, craftmanship and resourcefulness are evident in textile products throughout the ages; we are to this day awed by the beauty and sophistication of textiles sometimes found in archaeological excavations.

The objectives of the chemical processing, while basically unchanged over the centuries, have in recent times been diversified and expanded. Comfort and esthetics, durability and functionality, safety from fire and health hazards, easy care performance, such as washability, soil release, water and oil repellency, and stability against biological attack are examples of the objectives of chemical treatments of fibers and fabrics. Before these treatments can be applied, the textile materials have to be prepared by appropriate chemical procedures such as sizing, desizing, scouring, bleaching, and mercerization.

The array of fibers used at present is highly diversified. The advent of polyester, nylon, acrylic, and polyolefin fibers in recent years has greatly increased the complexity of the treatments as well as the range of the chemicals used. It became clear that approaches such as those practiced until 3 decades ago cannot continue to serve the solution to the wide range of problems facing chemists and technologists in the industry today. This realization coincided with rapid developments in polymer science and technology and brought about a surge in research and development activities in textile chemistry.

The studies carried out in the last 3 decades yielded a staggering amount of new data and not only a better understanding of the fibers and fiber assemblies and of the chemical interactions and structural changes, but also a large number of innovative ideas were created and put forward. Many of these ideas were developed into new processes, machines, and instruments, and culminated in a remarkable reshaping of the textile industry.

In these books an attempt is made to review and summarize the most important developments in this field. The emphasis is placed on the chemical aspects of the problems discussed. While technological aspects as well as industrial applications of the processes are being dealt with, only a brief treatment is given to factory layouts and to the machinery used.

Chemical Processing of Fibers and Fabrics is divided into two major areas. The first area, the fundamentals underlying the chemical treatments of fibers and fabrics and the preparation processes are presented in Vol. I, Parts A and B. The second area, the functional finishes of textiles are discussed in Vol. II, Parts A and B.

The need for a new comprehensive book in the field of chemical processing of fibers and fabrics has been felt for a long time. The vast amount of information accumulated in recent years in this field necessitated the preparation of the present books. They are intended for scientists and technologists both in the field of textiles and polymers as well as for students and researchers in other fields of human endeavor.

It is hoped that these books will not only further the knowledge and understanding of the complex field of textile chemistry, but will also bring about an interaction between people dealing in this field and people of other disciplines and will trigger off new and innovative developments for the benefit of all humanity.

Menachem Lewin
Stephen B. Sello

PREFACE

This is the first of two parts on fundamentals and preparation. It discusses several fundamental topics in the field of textile chemistry, such as interactions of fibers with water and organic solvents. Such interactions play an important role in all wet processes of fibers and fabrics which include preparation, dyeing, printing, and finishing. Of particular interest are the recent developments in the field of solvent treatments of fibers which may become of significant importance in the near future, from the point of view of saving energy.

While treatments of cellulosics with alkali have been carried out in the textile industry for over a century, it is still of great interest both to the scientist as well as to the technologist and new and unexpected developments are still taking place within this seemingly simple system. This is especially true for ammonia treatments which developed remarkably during the last decade. Two chapters deal with the effect of sodium hydroxide and liquid ammonia on fibers and fabrics and discuss their basic effect on morphology and crystal structure as well as the processing technologies, parameters and changes in physical properties and wear characteristics. The last chapter reviews the recent developments in the technology of raw wool scouring with emphasis on wool grease recovery and wastewater disposal system. Other preparation technologies such as warp sizing, desizing of textiles including size recovery systems, bleaching of wool and synthetics, and the application of optical whiteners are being reviewed in the second part which will be published in the not-too-distant future.

The editors wish to thank the editorial advisory board of the International Fiber Science and Technology Series, the contributors, and the editorial staff of Marcel Dekker for their cooperation and their contributions to this book.

Menachem Lewin
Stephen B. Sello

CONTRIBUTORS

Jean-Jacques Donzé Centre de Recherches Textiles de Mulhouse, Mulhouse, France

René Freytag École National Supérieure de Chimie de Mulhouse, Mulhouse, France

Luis G. Roldán-González Microscopy and Physics Department, Technical Center, J. P. Stevens & Co., Inc., Greenville, South Carolina

István Rusznák Department of Organic Chemical Technology, Technical University of Budapest, Budapest, Hungary

Catherine V. Stevens Sandoz Colors and Chemicals, East Hanover, New Jersey

Hans-Dietrich Weigmann Textile Research Institute, Princeton, New Jersey

George F. Wood* Commonwealth Scientific and Industrial Research Organization, Division of Textile Industry, Belmont, Victoria, Australia

---

*Now retired.

CONTENTS

About the Series iii
Introduction to the Handbook vii
Introduction to Volumes I and II ix
Preface xi
Contributors xiii
Contents of Other Volumes xix

1. Interactions Between Fibers and Organic Solvents 1

Hans-Dietrich Weigmann

1. Introduction 2
2. Thermodynamic Considerations in Fiber-Solvent Interactions 3
3. Diffusion of Organic Solvents into Polymeric Fibers 9
4. Solvent Induced Modification of Fiber Structure 15
5. Solvent Induced Shrinkage in Oriented Polymeric Systems 20
6. Solvent Induced Changes in the Physical Properties of Fibers 27
7. Solvents in Dyeing and Finishing Operations 34
8. Conclusion 44

References 44

2. Interaction of Aqueous Systems with Fibers and Fabrics 51

István Rusznák

1. Introduction 52
2. General Considerations 52
3. Sorption Processes in Textile Chemistry 55
4. Mass Transfer in Textile Chemical Processes 70
5. Penetration 72
References 91

3. Alkali Treatment of Cellulose Fibers 93

René Freytag and Jean-Jacques Donzé

1. Introduction 94
2. Action of Alkaline Agents on Cellulose Fibers 94
3. Scouring of Cotton 111
4. Mercerization of Cotton Fibers 134
References 157

4. Liquid Ammonia Treatment of Textiles 167

Catherine V. Stevens and Luis G. Roldán-González

1. Introduction 168
2. Properties of Liquid Ammonia 168
3. Effect of Liquid Ammonia on Structure and Morphology of Cotton 170
4. Effect of NH_3 on Rayon or Mercerized Cellulose 175
5. Liquid NH_3 Treatment of Cellulosics as a Substitute for Mercerization or as a Pretreatment for Easy Care Finishing 176
6. Liquid NH_3 in the Application of Topical Finishes 190
7. Dyeing from Liquid Ammonia 193
8. Liquid Ammonia Treatment of Noncellulosic Fibers 195
9. Equipment 198
10. Future of Liquid Ammonia Processing 198
References 199

5. Raw Wool Scouring, Wool Grease Recovery, and Scouring Wastewater Disposal 205

George F. Wood

1. Introduction 206
2. Detergent Scouring 213

3. Wool Grease Recovery 233
4. Wastewater Disposal 241
5. Energy Conservation 249
6. Solvent Scouring 250
7. Other Methods of Cleaning Raw Wool 253
References 254

Index 259

CONTENTS OF OTHER VOLUMES

Handbook of Fiber Science and Technology: Volume I
Chemical Processing of Fibers and Fabrics

FUNDAMENTALS AND PREPARATION PART B

Materials and Processes for Textile Warp Sizing, *Giuliana C. Tesoro and Peter Drexler*

Bleaching, *Menachem Lewin*

The Fluorescent Whitening of Textiles, *Raphael Levene and Menachem Lewin*

Wool Bleaching, *Raphael Levene*

Handbook of Fiber Science and Technology: Volume II
Chemical Processing of Fibers and Fabrics

FUNCTIONAL FINISHES PART A

Cross-Linking of Cellulosics, *Giuliana C. Tesoro*

Cross-Linking with Formaldehyde Containing Reactants, *Harro Petersen*

Finishing with Foam, *George M. Bryant and Andrew T. Walter*

Protection of Textiles from Biological Attack, *Tyrone L. Vigo*

FUNCTIONAL FINISHES PART B

Flame Retardance of Fabrics, *Menachem Lewin*

Repellent Finishes, *Erik Kissa*

Soil Release Finishes, *Erik Kissa*

Antistatic Treatments, *Stephen B. Sello and Catherine V. Stevens*

The Chemical Technology of Wool Finishing, *Trevor Shaw and Max A. White*

Radiation Processing, *William K. Walsh and Wadida Oraby*

Handbook of Fiber Science and Technology: Volume I

Chemical Processing of Fibers and Fabrics

FUNDAMENTALS AND PREPARATION

Part A

1

INTERACTIONS BETWEEN FIBERS AND ORGANIC SOLVENTS

HANS-DIETRICH WEIGMANN Textile Research Institute, Princeton, New Jersey

1. Introduction 2

2. Thermodynamic Considerations in Fiber-Solvent Interactions 3

2.1 Solubility parameter theory 4
2.2 Flory-Huggins theory 6
2.3 Solubility parameter principle and fiber-solvent interactions 7

3. Diffusion of Organic Solvents into Polymeric Fibers 9

3.1 General considerations 9
3.2 Fickian diffusion 10
3.3 Non-Fickian diffusion 13
3.4 Case II diffusion 14

4. Solvent Induced Modification of Fiber Structure 15

4.1 Effects of solvents on melting temperature 15
4.2 Effects of solvents on glass transition temperature 15
4.3 Solvent induced crystallization (SINC) 16
4.4 Surface and internal cavitation 17
4.5 Oligomer extraction by solvents 19

5. Solvent Induced Shrinkage in Oriented Polymeric Systems 20

5.1 General considerations 20
5.2 Isothermal shrinkage 20
5.3 Dynamic shrinkage 22
5.4 Solvent mixtures 26

6. Solvent Induced Changes in the Physical Properties of Fibers 27

6.1 Fiber structure-property relationships 27
6.2 Plasticization of polymeric fibers by solvents 28
6.3 Effects of solvents on fiber mechanical properties 30
6.4 Effects of solvents on dynamic mechanical properties of fibers 33

7. Solvents in Dyeing and Finishing Operations 34

7.1 Solvent dyeing 35
7.2 Solvent pretreatments 38
7.3 Solvent scouring and finishing processes 43

Conclusion 44

References 44

1. INTRODUCTION

In the last few decades the fiber-producing and textile industries have begun to realize that the discharge of more or less untreated wastes from processing, dyeing, and finishing plants is an unacceptable environmental hazard and that the era of cheap and plentiful water is irrevocably past. This realization and concurrent pressure from lawmakers and public opinion have resulted in serious efforts, not only to treat wastes to acceptable pollutant levels, but also to absorb the resulting costs through advanced technology. One approach that has been explored to a considerable extent and that initially has been enthusiastically embraced is the total substitution for water of a processing medium that can be recycled easily and inexpensively. A number of organic solvents offer themselves for this purpose, and various processes using such solvents have indeed been developed.

Initially, the most successful solvent process was solvent scouring, the use of supposedly noninteractive solvents to eliminate undesirable substances from the surface of the fiber. It was soon discovered, however, that solvents, water among them, invariably interact with the substrate, although the degree of interaction varies depending on the nature of the solvent, the nature of the fiber substrate, and the temperature of the process. The thermodynamics of the interaction between solvents and polymeric fibers and the kinetics of diffusion of the solvents into the substrate have to be taken into consideration.

Because the cost of energy has become increasingly prohibitive, energy-efficient processes involving interactive media that permit dyeing and finishing at lower temperatures have come under close scrutiny. It has been found that solvent treatments result in structural modifications of polymeric fibers which open the structure sufficiently to facilitate subsequent conventional dyeing and finishing operations. The following sections deal with the thermodynamics of the interactions of solvents with polymeric substrates, the diffusion processes of solvents in polymeric fibers, and the resulting modifications of the structure and properties of the fiber substrate. Finally, a brief section on the application of the fiber-solvent interaction principle in industrial processes will be presented. Reference should be made at this point to comprehensive review papers which have appeared in the recent literature covering the interactions of solvents with polymers in general [1] and with fibers in particular [2].

2. THERMODYNAMIC CONSIDERATIONS IN FIBER-SOLVENT INTERACTIONS

The extent of interaction between polymeric fibers and interactive solvents is strongly influenced by morphological and structural considerations. In the simple case of an amorphous noncrystallizable polymer, the solvent can quite readily enter the polymer, replace polymer-polymer interactions with polymer-solvent interactions, and cause more or less extensive swelling up to complete solubilization. In semicrystalline, oriented polymers as they exist in textile fibers, segmental motion is inhibited and network chain separation is limited. Fiber-solvent interactions therefore manifest themselves in the absorption-swelling equilibrium rather than in solubilization. The same statement can be made for cross-linked polymers, which are virtually impossible to solubilize because of their extremely high molecular weight. The extent of swelling in a solvent depends on the balance between the osmotic pressure of the solvent and the retractive elastic forces of the polymer network. Semicrystalline, oriented artificial fibers behave similarly to cross-linked materials, provided the crystallites are large enough to act as physical cross-links between amorphous domains. Solubilization is possible only if the extent of interaction is strong enough to dissolve or "melt" the crystallites.

Aside from morphological factors, the extent of interaction between polymer and solvent is determined by the nature of the forces between polymer molecules and the ability of the solvent specifically to break these interactive forces and replace them with polymer-solvent interactions. The interactive forces in nonelectrolytic polymers that have to be considered in this context are dispersion (London) forces, polar forces, induction forces, and hydrogen bonding. Dispersion forces are operative between all molecules, and they are the only kind of interaction between nonpolar materials. These forces are caused by nonpermanent, fluctuating dipoles which produce strongly attractive, nondirectional forces between polymer molecules. Polar forces, on the other hand, originate from permanent dipoles existing in polar groups of the polymer unit, which can interact with other permanent dipoles to produce electrostatic interactions. The strength of these polar forces depends on the orientation of the dipoles. Induction forces exist where dipole-containing groups sufficiently polarize another molecule to induce a dipole, resulting in attractive forces similar to polar forces. A much stronger interaction than any of the preceding is hydrogen bonding. In this kind of interaction, a hydrogen bridge is formed through the sharing of a hydrogen atom between two strongly electron-withdrawing atoms, such as oxygen or nitrogen.

2.1 Solubility Parameter Theory

The interactions between polymer and solvent can be described by the change in free energy ΔG_m which occurs during mixing:

$$\Delta G_m = \Delta H_m - T\Delta S_m \tag{1}$$

where ΔH_m is the enthalpy change and ΔS_m is the entropy change upon mixing. This change in free energy must be negative for the process to proceed spontaneously. A change in entropy during mixing of polymer-solvent systems is always positive [3-5], since the separation of polymer chains by solvent molcules increases segment mobility and with it randomness of the polymer-solvent system. Thus, the negativity of the overall change in free energy of mixing for the polymer-solvent system depends upon the magnitude of the enthalpy term ΔH_m.

The most widely accepted expression for this term was introduced by Hildebrand and Scott [6] and by Scatchard [7]:

$$\Delta H_m = V_m \left[\left(\frac{\Delta E_1}{V_1} \right)^{1/2} - \left(\frac{\Delta E_2}{V_2} \right)^{1/2} \right]^2 \phi_1 \phi_2 \tag{2}$$

where V_m is the total volume of the mixture of two compounds with energies of evaporization ΔE and molar volumes V, and ϕ_1 and ϕ_2 are the volume fractions of the two compounds in the mixture. The ratio $\Delta E/V$ has been called the *cohesive energy density* since it is a measure of the energy required to overcome all the molecular forces in 1 mol of the substance. Hildebrand has introduced the solubility parameter δ defined as the square root of the cohesive energy density,

$$\delta = \left(\frac{\Delta E}{V} \right)^{1/2} \tag{3}$$

It is apparent from Eq. (2) that the heat-of-mixing term approaches 0 when the solubility parameters of solvent and polymer are equal. Under these conditions, the solvent would be expected to solubilize the polymer.

A number of assumptions have been made in the development of

Hildebrand's heat-of-mixing term; these have been reviewed by Gardon [5,8]. Because these assumptions are not always all valid, there are occasional discrepancies between theory and experimental fact, especially in the rare cases of exothermic mixing, which is not possible according to the Scatchard-Hildebrand theory. It has been shown, however, that both the Scatchard-Hildebrand theory and more elaborate theories of solution that avoid the simplifying assumptions [9] predict that a solvent for a given polymer is one whose solubility parameter is equal or close to that of the polymer. When expressed in terms of partial molar quantities, Eq. (2) can be reformulated as follows [6]:

$$\Delta\bar{E}_1^{\,m} = \phi_2^{\,2} V_1(\delta_1 - \delta_2)^2 \tag{4}$$

where $\Delta\bar{E}_1{}^m$ is the partial molar energy of mixing of the solvent and V_1 is the molar volume of the solvent. It can be concluded from this equation that solubility is enhanced when the solvent has a small molar volume, a tendency that has indeed been observed [10-12].

The solubility parameter theory can be extended to include solvent mixtures [3,13]. The solubility parameter δ_{mix} of a solvent mixture can be expressed

$$\delta_{mix} = \frac{x_1 V_1 \delta_1 + x_2 V_2 \delta_2}{x_1 V_1 + x_2 V_2} \tag{5}$$

where x, V, and δ stand for the mol fraction, the molar volume, and the solubility parameter of the components of the mixture, respectively. This expression becomes invalid when there is a change in volume on mixing.

When attempts are made to correlate solubility characteristics of polymer-solvent systems with their solubility parameters, anomalies are frequently encountered, especially when one is dealing with highly polar or hydrogen bonding systems. It became necessary, therefore, to define additional parameters to characterize the polarities and the hydrogen bonding capabilities of solvents and polymers. A number of investigators have attempted to separate polar and nonpolar interactions and to establish the contributions of the various components to the total solubility characteristics of polymer-solvent systems. Perhaps the most successful approach is that of Hansen [10,14], who introduced the three-dimensional solubility parameter concept, in which the total cohesive energy of a material is divided into dispersion d, polar p, and hydrogen bonding h contributions. The total solubility parameter is then expressed

$$\delta = (\delta_d^{\,2} + \delta_p^{\,2} + \delta_h^{\,2})^{1/2} \tag{6}$$

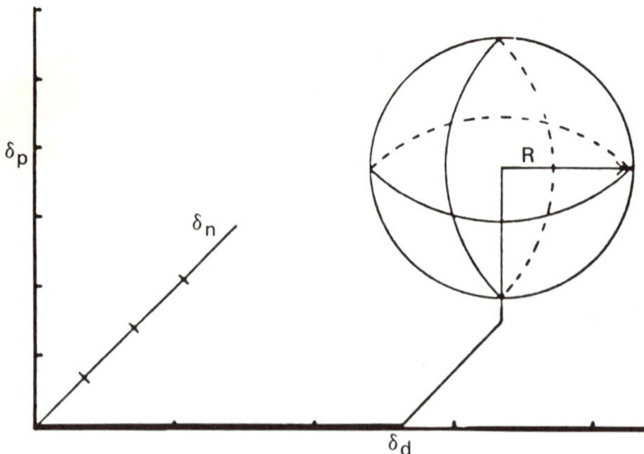

Figure 1.1 Sketch of typical volume of interaction for Hansen's three-dimensional solubility parameter concept. (*Source*: From Hansen [10] by permission of the publisher, The American Chemical Society, Copyright 1969.)

Hansen was able to establish the component solubility parameters for many solvents and polymers and was able to make useful predictions concerning miscibility by finding spheres of solubility in three-dimensional plots such as that in Fig. 1.1. In these plots the polymer is represented by a point determined by its coordinates ($\delta_{d,P}$, $\delta_{p,P}$, $\delta_{h,P}$). Each solvent, also defined by its coordinates ($\delta_{d,S}$, $\delta_{p,S}$, $\delta_{h,S}$), that lies within the sphere is a true solvent for the polymer, while liquids lying outside the sphere show only partial or no miscibility with the polymer. The radius of the sphere is found experimentally in numerous solubility tests. Hansen's composite solubility parameters have been listed [5,10] and have been supported by independent theoretical calculations [14]. Using Hansen's three-dimensional solubility parameter values, the maxima of interaction between various polymers and binary and ternary solvent mixtures have been calculated [15-17].

2.2 Flory-Huggins Theory

This theory represents an alternative approach to quantitative description of polymer-solvent interactions [18-20]. The free energy of mixing between polymer and solvent can be represented by

$$\frac{\Delta G_m}{RT} = n_1 \ln \phi_1 + n_2 \ln \phi_2 + \chi\phi_1\phi_2(n_1 + mn_2) \tag{7}$$

where n_1 and n_2 are the number of mols of solvent and polymer, respectively, m is the ratio of polymer to solvent molar volume, and χ is the so-called Flory interaction parameter, which is a measure of the interaction energy between any given solvent and solute. Huggins has calculated the enthalpy term of the χ parameter using assumptions of the Scatchard-Hildebrand theory and has arrived at the expression

$$\chi_H = \frac{V_S}{RT}(\delta_P - \delta_S)^2 \qquad (8)$$

where the subscripts P and S refer to polymer and solvent, respectively. The Flory-Huggins theory predicts that dissolution of a polymer by a particular solvent occurs only if the χ value of the system is less than a critical value χ_c given by the equation

$$\chi_c = \frac{1}{2}\left(1 + \frac{1}{M^{1/2}}\right)^2 \qquad (9)$$

so that χ_c for very high molecular weight (M) polymers has a value of 0.5. The application of the Flory-Huggins theory to predict polymer-solvent interactions has met with limited success, and it appears that the solubility parameter concept, once the values for the composite parameters have been determined, is more powerful and simpler in its application.

2.3 Solubility Parameter Principle and Fiber-Solvent Interactions

In attempts at quantitative description of polymer-solvent interactions involving semicrystalline polymers, it must be considered that the potential energy difference for polymer segments moving from the ordered crystalline state to the solvated state is much higher than from the amorphous to the solvated state. The heat of fusion of the crystallites has to be included in the free-energy equation, but no sound theoretical approach has as yet been discussed for this case. Hildebrand [6] has suggested that crystallites in the amorphous region be treated as separate components because of the relative impenetrability of the polymer crystallites. Even good solvents penetrate and swell only the amorphous domains, leaving the crystallites essentially unaffected.

The application of the solubility parameter principle to quantify or predict polymer-solvent interactions has been attempted in a number of cases. Moore and Sheldon [21] have correlated the ability of solvents to swell and crystallize unoriented amorphous polyethylene terephthalate (PET) with the total solubility parameter of the solvent. These authors observed a bimodal distribution when plotting swelling or density of the solvent treated polymer against the solubility

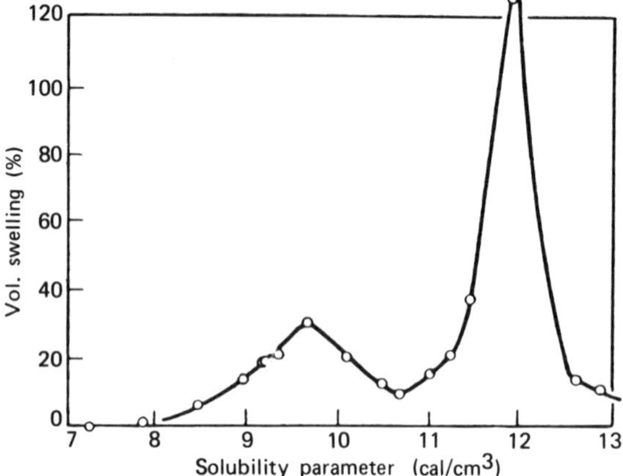

Figure 1.2 Equilibrium swelling of the polymer after liquid penetration as a function of the solubility parameter of the liquid. (*Source*: From Moore and Sheldon [21] by permission of the publishers, IPC Business Press Ltd. ©)

parameter of the solvent (Figs. 1.2 and 1.3). A similar distribution is observed when solvent induced crystallinity data obtained by Lawton and Cates [22] are plotted against the solubility parameter of the solvent [23]. Weigmann and co-workers [23-27] and Kimihiro Suzuki [28] have studied the mechanical properties, swelling, and shrinkage behavior of oriented semicrystalline PET fibers and found bimodal

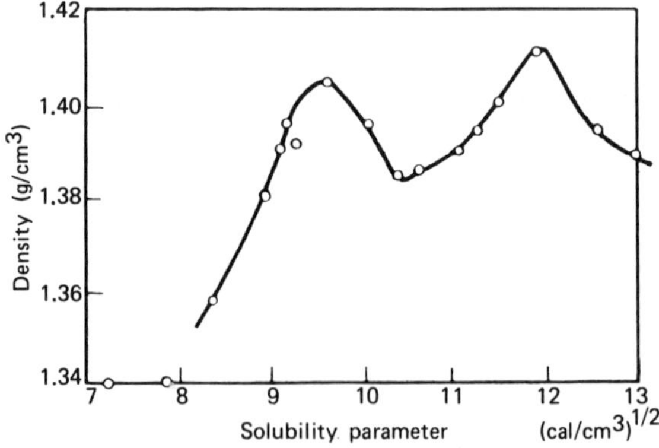

Figure 1.3 Equilibrium density of the polymer after liquid penetration as a function of the solubility parameter of the liquid. (*Source*: From Moore and Sheldon [21].)

distributions for various properties reminiscent of the data of Sheldon et al.

In all these examples it was found that maximum interactions with the polymer occur at two regions of solubility parameter values: $\delta_1 = 9.8$ and $\delta_2 = 12.1$. It is assumed either that in these regions solvents interact preferentially with the aromatic or the aliphatic residue of the PET repeat unit, or that there are specific acid-base interactions between solvent and polymer groups. The polymer can be considered an alternating copolymer such that a given solvent can interact with one or the other of the repeating segments. Knox [29] has explored this aspect in considerable detail, also finding two interaction ranges when he applied the component solubility parameters of Hansen to the interactions of solvents with PET.

Ingamells and co-workers [30] applied component solubility parameters to an analysis of the reduction of the glass transition temperature T_g of PET by solvents. These authors used the approach of Blanks and Prausnitz [31] in which the total solubility parameter is represented by two components associated with the dispersion and polar forces, referred to as association forces. Ingamells and coworkers [32,33] claimed earlier that dispersion forces are dominant in the interactions between PET and solvents. Two values of δ_d were found for PET, and the authors suggest that this observation supports the claim that aromatic and cyclic solvents solvate the aromatic ester residue, while aliphatic solvents interact with the aliphatic ester residue of the repeat unit. The lowering of the T_g of polyacrylonitrile fibers by solvents absorbed from an aqueous environment, on the other hand, has been correlated with the value of the association solubility parameter [34]; no correlations with the dispersion component could be found. Lindner [35] has made measurements of the isothermal shrinkage behavior of PET filaments in organic solvents and has used the induction period as a means of determining diffusion coefficients. The apparent activation energy of solvent diffusion is strongly affected by polymer-solvent interactions, and correlations were established with the component solubility parameters of a number of solvents. Lindner used these correlations to calculate the component solubility parameter of the polymer. The total solubility parameter of PET was found to be 9.6 $(cal/cm^3)^{1/2}$, in contrast to a value of 10.7 reported by Moore [36]. The hydrogen bonding, polar, and nonpolar components were determined to be 4.5, 3.1, and 7.9 $(cal/cm^3)^{1/2}$, respectively.

3. DIFFUSION OF ORGANIC SOLVENTS INTO POLYMERIC FIBERS

3.1 General Considerations

Small penetrant molecules having low levels of thermodynamic interaction with the polymer move within the polymer structure by activated jumps from one "hole" to another. These holes cannot be considered to be actual voids or microvoids within the structure, especially at temperatures above T_g, as pointed out by Peterlin [37]. The diffusion of such small, weakly interactive penetrants is characterized by concentration-independent diffusion coefficients, and the minimum hole size required for penetrant jumps is smaller than the average free volume of the polymer under the experimental conditions.

The diffusion process is much more complex for penetrant molecules that are larger or have strong interactions with the polymer. The transport of larger molecules is possible only if there is rearrangement of polymer segments to permit the penetrant molecule to move in accordance with its concentration gradient within the polymer structure. Anomalous or non-Fickian mass transport behavior is frequently observed for solvents with strong thermodynamic interactions with the polymer. This behavior is attributed to relaxation responses of the polymer segments to swelling stresses generated as the solvent penetrates the polymer structure. As highly interactive solvents diffuse into the polymer, increased mobility of the polymer segments results, and in some cases large-scale structural rearrangements are observed to follow immediately behind the diffusion front, which can lead to solvent induced crystallization (SINC) of an initially amorphous polymer. This aspect will be discussed in more detail in a subsequent section.

3.2 Fickian Diffusion

The diffusion of organic penetrants into glassy polymers has been discussed extensively in the literature and has been reviewed recently [37]. A mathematical description of diffusion processes is given by Fick's two laws, the first of which states that the flux J of a penetrant is proportional to its concentration gradient $\partial c/\partial x$ normal to the unit area of cross section:

$$J = -D \frac{\partial c}{\partial x} \tag{10}$$

where D is the diffusion coefficient. The rate of change of concentration of the penetrant as a result of diffusion is given by Fick's second law:

$$\frac{\partial c}{\partial t} = D \frac{\partial^2 c}{\partial x^2} \tag{11}$$

if diffusion is unidimensional. The general case of Fick's second law can be expressed in the nomenclature of vector analysis as follows:

$$\frac{\partial c}{\partial t} = \text{div}\ [D(c)\ \text{grad}\ (c)] \tag{12}$$

which takes into account that the diffusion coefficient may be a function of the concentration of the penetrant. If the penetrant molecule in the polymer is smaller than the average free volume in the polymer, D is independent of concentration. However, in most cases where the penetrant molecule is larger than the average free volume or where specific interactions occur between penetrant and polymer, the diffusion coefficient exhibits concentration dependence. Kwei and Wang [38] have shown that the concentration dependence of the solvent diffusion coefficient in a particular polymer-solvent system depends entirely on the mobility of chain segments and therefore on the structure and morphology of the polymer.

Not only is the concentration dependence of the diffusion coefficient affected by the structural properties, but, in general, any decrease in segmental mobility in the polymer also decreases the rate of diffusion. Factors such as crystallization, orientation, cross-linking, chain stiffness, and so on, all decrease the rate of mass transport in the polymer [39,40]. Since polymeric fibers usually have a semicrystalline structure with significant degrees of orientation, some of these aspects will be discussed here.

Peterlin [37] has recently reviewed transport phenomena in semicrystalline polymeric materials and characterized contributions from the crystalline domains by means of a blocking factor B and an immobilization factor ψ

$$D = D_a \frac{\psi}{B} \tag{13}$$

where D_a is the diffusion coefficient for the completely amorphous polymer. Diffusive transport is assumed to take place almost exclusively through the amorphous domains. The diffusion paths are blocked by impenetrable crystalline lamellae, while the anchoring of the amorphous chains between crystallites and the resulting reduction in chain mobility is described by the immobilization factor. Peterlin based his discussions on the free-volume theory (see also [41]) and on a two-phase semicrystalline system in which the fractional free volume in the crystallites is so small that penetration into these domains is negligible. Uniaxial elastic strain on semicrystalline polymers increases the diffusion coefficient by increasing the fractional free volume of the amorphous regions [42]. An elastic strain normal to the crystalline lamellae increases the thickness of the amorphous layer between the lamellae, resulting in an increase in specific volume. Uniaxial plastic deformation of a spherulitic polymer, on the other hand, increases the orientation of the amorphous regions, resulting in a closer packing of the polymer chains parallel to the stretch direction. Crystallite deformation and parallelization of the amorphous chains produce a higher density and with it a lower fractional free volume and a decreased diffusion coefficient [43]. At very high elongation, the

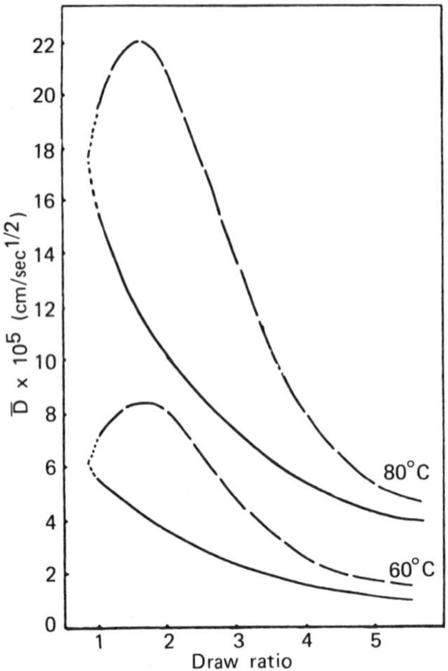

Figure 1.4 Radial ($D_\perp$ --) and axial ($D_\parallel$ —) diffusion coefficients of water in nylon 6 at 60 and 80°C as functions of the draw ratio. (*Source*: From Takagi [45] with permission of the publisher, John Wiley and Sons, Inc.)

crystallites themselves become completely disrupted and the crystalline texture changes from spherulitic to fibrous [44].

The drawing of polymeric fibers has a similar effect on diffusion coefficients. The diffusion coefficient perpendicular to the draw direction is significantly larger than that parallel to it [45], reflecting the higher probability of finding holes and channels in the structure in the radial direction. Diffusion paths in the axial direction can occur only along microfibrillar boundaries, since the diffusion paths within the microfibrils are blocked by lamellae. Takagi [45] studied the diffusion of water in nylon 6 fibers and observed an increase in the radial diffusion coefficient at low draw ratios, while the diffusion coefficient in the axial direction decreased (Fig. 1.4). Apparently the predominant effect at small elongations is the pulling apart and separation of lamellae oriented perpendicular to the stretch direction. At higher elongations, on the other hand, the lamellae begin to be transformed into microfibrillar material [44], and the plastic deformation of the structure produces higher density in the amorphous domain, thus decreasing diffusion coefficients in both radial and axial directions.

3.3 Non-Fickian Diffusion

While Fickian diffusion is observed in most polymeric materials at temperatures more than 15°C above their glass transition temperature, anomalous diffusion behavior is observed in certain cross-linked or crystalline polymers [46] and certainly in many polymers below the glass transition temperature. Crank and Park [47] noted that under certain conditions significant deviations are seen from the linear relationship between absorption and the square root of time predicted for Fickian behavior. These authors offered a number of hypotheses to explain this anomalous diffusion behavior, most of them involving time-dependent polymer behavior, i.e., relaxation phenomena in the glassy state having a slow time response. It is therefore unlikely that the introduction of a penetrant, especially an interacting one, into the polymer structure would produce anything but a time-dependent response. For instance, concentration equilibria in the polymer are

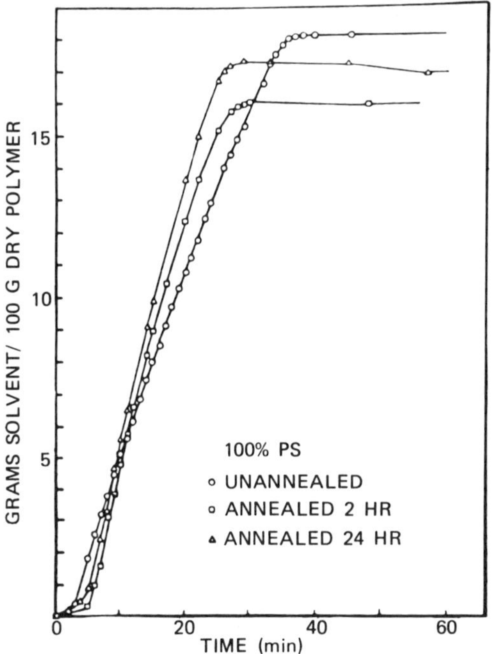

Figure 1.5 Kinetics of n-hexane vapor sorption at 40°C and 0.98 activity (ratio of n-hexane partial pressure to equilibrium partial pressure) in annealed and unannealed films of 100% glassy polystyrene. The films were annealed at 20°C above their glass transition temperature. (*Source*: From Hopfenberg, Stannett, and Folk [49] with permission of the publisher, Society of Plastic Engineers.)

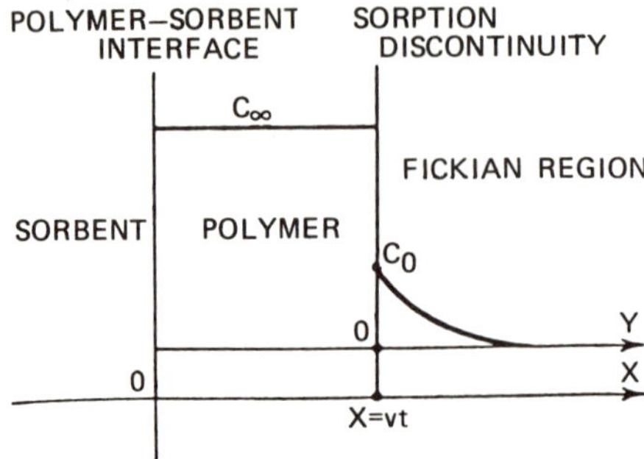

Figure 1.6 Schematic representation of the concentration distribution in a polymer ($X > 0$) with superimposed Fickian and case II diffusion: the Fickian region with $C < C_0(X > vt,\ y > 0)$, the sorption discontinuity at $X = vt$, $y = 0$, and the case II region $C_\infty > C_0(0 \leqq X < vt)$. v is the velocity of the advancing sorption discontinuity and y is a new coordinate system with its origin at the sorption discontinuity and moving with velocity v. (*Source*: From Peterlin [54] with permission of the publishers, Hüthig & Wepf Verlag, Basel.)

achieved not instantaneously but rather at a finite rate, leading to the conclusion that a time-dependent diffusion coefficient exists. It is also possible that the value of the diffusion coefficient at any given concentration may never really attain its true equilibrium value but merely approach it. Another hypothesis concerns the effects of swelling stresses at the penetrant front. The penetrated and the nonpenetrated regions exert high stresses onto each other which directly affect the diffusion coefficient. As diffusion proceeds and the unpenetrated, unswollen region becomes smaller, the diffusion coefficient increases until it finally reaches its stress-independent value.

3.4 Case II Diffusion

Alfrey and co-workers [48] have discussed an extreme case of non-Fickian diffusion behavior that manifests itself in the following way: A sharp penetration front separates the rubbery, penetrated outer domain with a uniform penetrant concentration from the inner, glassy core containing essentially zero penetrant. This penetration front advances at constant velocity into the polymer, and the weight gain of the polymer is linear with time over a considerable penetration range (Fig. 1.5). Most investigators agree that the rate-controlling processes in case II diffusion are due to the relaxation of stresses produced during swelling [50-54]. A number of workers have shown that case II and Fickian diffusion can be superimposed by varying the cross-link density of the polymer or by varying the type of penetrant [50, 51, 54]. A schematic representation of the concentration distribution in a polymer with superimposed Fickian and case II diffusion is shown in Fig. 1.6. Recently, Haga [55] has established relationships between case II diffusion of solvents into PET and the microstructure of the polymer.

4. SOLVENT INDUCED MODIFICATION OF FIBER STRUCTURE

4.1 Effects of Solvents on Melting Temperature

Both the melting and glass transition temperatures of a polymer can be depressed by the presence of a solvent, and the degree of this depression depends on the amount of solvent and its interactions with the polymer. Flory [56] has presented a quantitative relationship between the melting point T_m of a semicrystalline polymer and the amount of diluent in the amorphous phase:

$$\frac{1}{T_m} - \frac{1}{T_m^\circ} = \frac{RV_2}{\Delta H_2 V_1}(\phi_1 - \chi_1\phi_1^2) \tag{14}$$

where

T_m° = melting temperature without diluent

V_1 and V_2 = molar volumes of diluent and polymer repeat unit, respectively

χ_1 = Flory-Huggins interaction parameter

ϕ_1 = volume fraction of diluent

ΔH_2 = heat of fusion per mol of repeat unit of polymer

This relationship is valid provided the diluent does not penetrate the crystalline phase, an assumption that appears to be acceptable for most semicrystalline polymeric fibers.

4.2 Effects of Solvents on Glass Transition Temperature

A number of equations have been derived to describe quantitatively the lowering of the polymer glass transition temperature in the presence of a diluent. The assumption is usually made that the diluent or solvent acts as a plasticizer for the noncrystalline domains, breaking intermolecular bonding and thereby providing a lubrication action resulting in enhanced mobility of the polymer segments. Solubility of the solvent in the polymer is an obvious requirement, and this compatibility determines the equilibrium uptake of the solvent by the polymer and with it the extent of T_g depression. One of the more frequently used relationships between T_g depression and diluent uptake has been derived by Gordon and Taylor [57]:

$$T_g = \frac{KW_1T_{g_1} + W_2T_{g_2}}{KW_1 + W_2} \qquad (15)$$

where 1 and 2 represent plasticizer and polymer, respectively, and W_1 and W_2 are the weight fractions of the two components. The parameter K is determined by the difference between the expansion coefficients β of the melt and glass of the two components:

$$K = \frac{\Delta\beta_2}{\Delta\beta_1} \qquad (16)$$

If volume fractions are used instead of weight fractions in Eq. (15), the relationship is referred to as the Kelley-Bueche equation. It is important to point out that the extent of polymer-solvent interaction appears in Flory's equation for T_m depression through the χ term, while such a term is absent in the Gordon-Taylor equation.

4.3 Solvent Induced Crystallization (SINC)

The glass transition temperature of a crystallizable polymer may be sufficiently depressed by the presence of a solvent that segmental mobility becomes sufficient to permit crystallization. An important difference between SINC and purely thermal crystallization is that, in the case of amorphous polymers and even in the case of semicrystalline polymers such as oriented fibers, crystallization takes place in the presence of another molecular species, the solvent, and therefore in a swollen state of the polymer.

The superstructure produced by SINC in initially unoriented, amorphous polymeric materials has been investigated by various workers [58-60]. Balcerzyk [61] has concluded from the crystallization kinetics of amorphous PET fibers in alcohol solutions of methylene chloride that the SINC process is spherulitic in character. Using optical microscopy, small-angle light scattering, and scanning electron microscopy, Wilkes and his co-workers [60] were able to confirm directly that spherulitic structures are characteristic of the morphology induced by solvent treatments in unoriented amorphous polymers (Fig. 1.7). These authors have also shown that spherulitic crystallization occurs during the absorption process, rather than upon the removal of the solvent from the polymer as had been suggested previously [62]. Wide-angle x-ray scattering patterns of solvent crystallized PET films or fibers do not show sharp reflections, even though crystallinity values determined by density are on the order of 40-50%. These findings clearly indicate that high concentrations of small crystallites are formed during solvent induced crystallization, reflecting the very high nucleation rate of the polymer in the presence of the solvent [27,63-67].

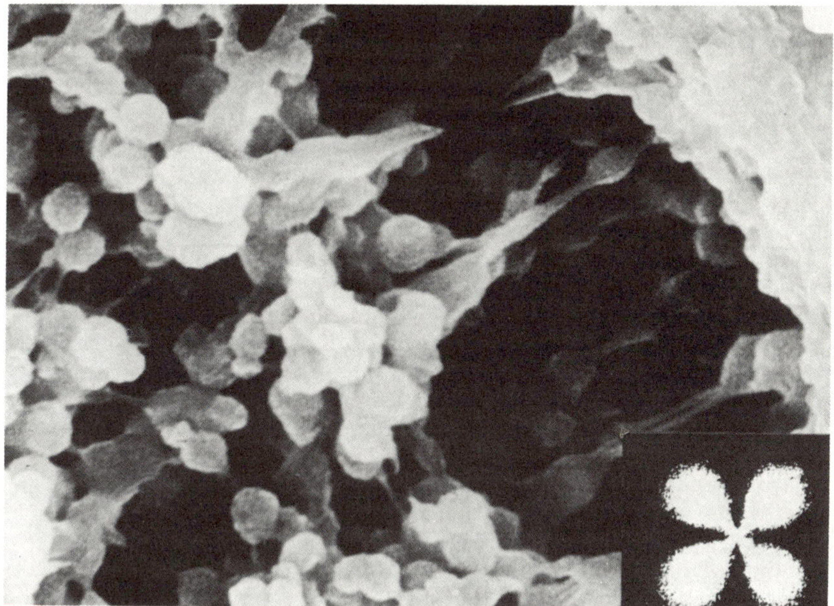

Figure 1.7 Typical spherulitic surface morphology and corresponding H_VSALS pattern of an amorphous PET film treated with dioxane at 26°C for less than 1 min. (*Source*: From Desai and Wilkes [60] by permission of the publisher, John Wiley and Sons, Inc.)

4.4 Surface and Internal Cavitation

Exposure of an unoriented amorphous film or fiber to an interactive, crystallizing solvent may produce extensive cavitation of the polymer surface. It is assumed [60,64] that this cavitation arises from a combination of anisotropic swelling forces, solvent stress cracking or crazing, and spherulite formation caused by crystallization which leads to localized densification. Surface cavitation is not observed during vapor sorption or in oriented (cold-drawn) PET films or fibers, even though solvent absorption and crystallization do occur. The depth of cavitation depends on the SINC temperature, contact time with the solvent, and the nature of the solvent, but never appears to exceed the order of 100 μm.

Solvent induced surface cavitation has been utilized by Weigmann and co-workers [68,69] to produce bimorphic PET fibers with extremely high surface/volume ratios (Fig. 1.8). Partial penetration of the solvent into the surface domains of the fiber (Fig. 1.9) produces a highly cavitated layer of solvent crystallized polymer. After removal of the solvent, conventional structure development in the unpenetrated core is achieved by stress-induced crystallization during high-temperature drawing, resulting in fibers of adequate overall mechanical properties. It is possible to form these fibers in a continuous process of sequential solvent immersion and drawing. The spherulitic surface domains are deformed during the drawing step to form an oriented

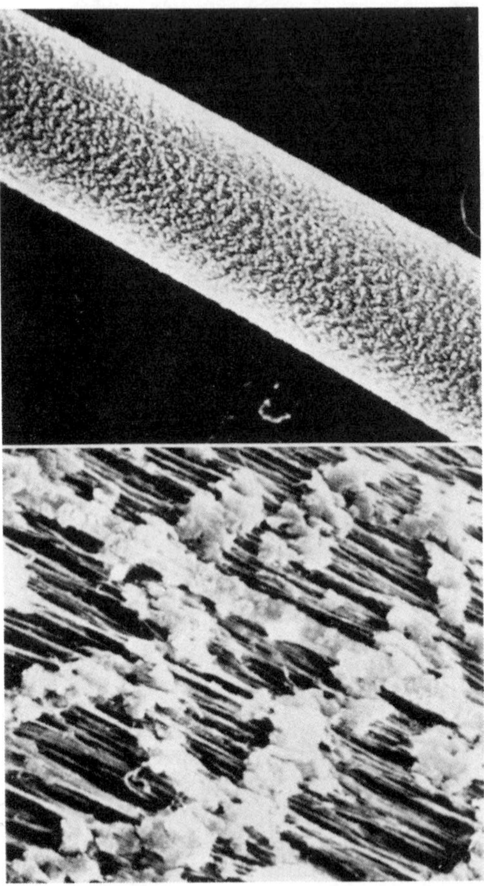

Figure 1.8 SEM micrographs of monofilament yarn samples drawn through DMF and a waterbath rinse at DR 4; HPT = 100°C. a = 130X; b = 1900X. (*Source*: From Gerold et al. [68] with permission of the publisher, Textile Research Institute.)

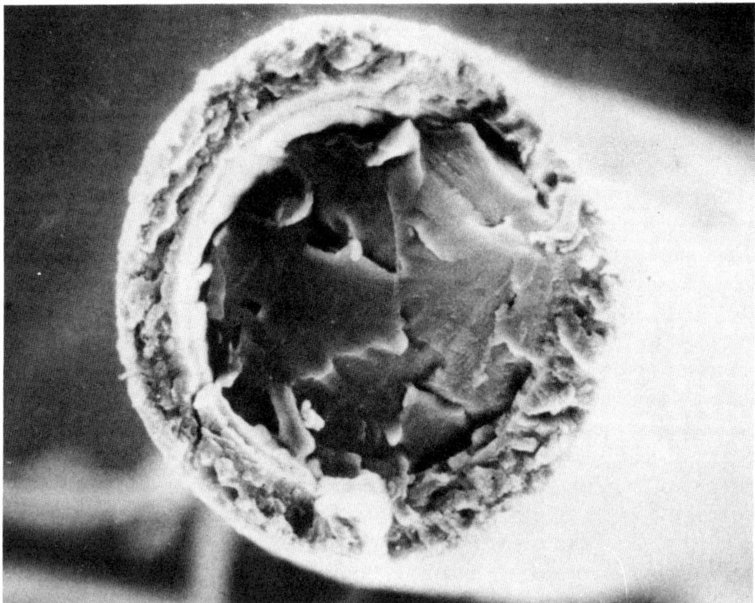

Figure 1.9 SEM brittle fracture micrographs of monofilament yarn surface samples drawn at DR 4 through DMF; HPT = 100°C. (*Source*: From Gerold et al. [68].)

fibrillar structure of some strength and stability, while the high surface/volume ratio is retained. The usefulness of the high surface area in various applications, such as improved adhesion in fiber reinforced composites, moisture transport in apparel use, and the decomposition of appropriate materials in the surface cavities for slow release, is currently under investigation.

4.5 Oligomer Extraction by Solvents

Under certain conditions of solvent exposure or at elevated temperatures, low-molecular-weight species or oligomers, which originally are dispersed throughout the bulk polymer, migrate to the surface of fibers and crystallize. Their presence at the fiber surface and their accumulation on textile-processing equipment often produce undesirable effects. Certain solvents are capable of extracting these oligomers, and Bredereck and co-workers [70] have systematically investigated oligomer extraction from PET fibers in a number of solvents and studied the kinetics of the extraction process.

5. SOLVENT INDUCED SHRINKAGE IN ORIENTED POLYMERIC SYSTEMS

5.1 General Considerations

Shrinkage of oriented polymeric systems occurs as a result of molecular relaxation of orientational strains in the direction of the principal orientation axis. Molecular relaxation of this nature involves substantial movement of polymer segments and requires temperatures above the glass transition temperature. This can be achieved either by raising the temperature of the polymer above its glass transition temperature or by lowering the glass transition temperature through interactions with interactive solvents.

Thermally induced shrinkage of oriented semicrystalline polyamide filaments [71-73] and polyester filaments [74,75] has been discussed both in thermodynamic and in kinetic terms. The initial relaxation and disorientation during shrinkage is frequently followed by melting and recrystallization of imperfect crystalline domains or crystallization of noncrystalline domains, normally as a secondary step after shrinkage [76]. Shrinkage either can be prevented completely or can be limited by placing the fibers under mechanical constraint during the heating process. In this case, shrinkage forces will be generated reflecting the molecular processes of relaxation and crystallization. If extensive viscoelastic stress relaxation or crystallization occurs under these constant-length treatment conditions, the fibers are set; i.e., subsequent heating under unconstrained conditions will not result in shrinkage or at least will result in lower shrinkage levels. Certain materials with an extremely high tendency toward crystallization do not shrink until temperatures just below their crystalline melting point are reached. A typical example of this behavior is oriented isotactic polypropylene [77]. As was pointed out above, interactive solvents can lower the glass transition temperature of a dry polymer substantially, resulting in considerable levels of solvent induced shrinkage, provided the cross section of the filaments has been completely penetrated.

5.2 Isothermal Shrinkage

The interactions of oriented semicrystalline polyamide filaments with aqueous phenol solutions was one of the first systems to be investigated in detail in terms of solvent induced shrinkage behavior. The effects of phenol concentration [78] and draw ratio have been studied. Jacobs et al. [79] have investigated the kinetics of phenol induced shrinkage and interpreted their results in terms of two sequential first-order processes (constants k_1 and k_2) involving intermolecular bond breakdown and subsequent bond reformation. The shrinkage kinetics can be expressed by the following equation:

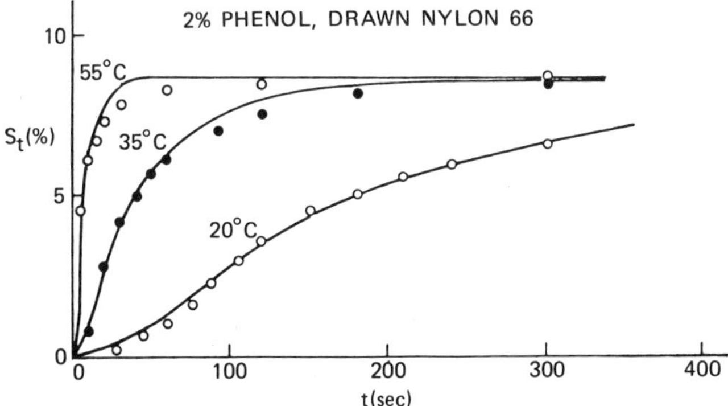

Figure 1.10 Shrinkage as a function of time for drawn nylon 66 filaments in 2% aqueous phenol solution. Lines are calculated from Eq. (17); points represent experimental measurements. (*Source*: From Jacobs et al. [79] with permission of the publisher, John Wiley and Sons, Inc.)

$$\frac{S_t}{S_\infty} = \frac{k_2[1 - \exp(-k_1 t)] - k_1}{k_2 - k_1} \tag{17}$$

which describes the experimental data satisfactorily (Fig. 1.10). It was concluded that hydrophobic interactions as well as hydrogen bonding between polyamide chains contribute to the stability of the nylon 66 structure [80].

The kinetics of thermal and solvent induced shrinkage of polyester filaments has been studied by Weigmann and co-workers [24], using a broad range of solvents. It was found that the rate of shrinkage of drawn polyester filaments depends to a large extent on the nature of the solvent, i.e., both its molecular size, which influences its ability to diffuse into the fiber structure, and its extent of interaction, which determines the degree of lowering of the T_g of the polymer. The maximum rate of shrinkage has been used to characterize the shrinkage process, and it appears that shrinkage is a first-order process, the temperature dependence of which obeys a simple Arrhenius relationship (Fig. 1.11). Moore et al. [81] had previously shown that both thermal and solvent induced shrinkage follow an Arrhenius relationship. The activation energy of the solvent induced process is substantially higher than that for thermal shrinkage, suggesting that solvent induced shrinkage involves movement of solvated polymer segments of significantly larger volume and requires higher energy.

The induction time, i.e., the time necessary for shrinkage to begin, has been used to characterize the rate of solvent diffusion

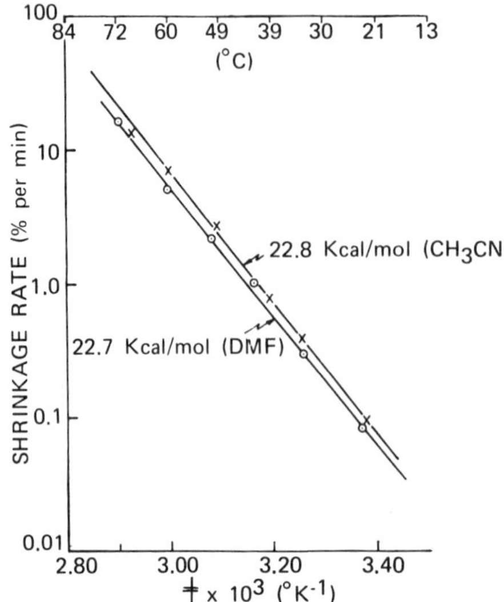

Figure 1.11 Arrhenius-type temperature dependence of maximum shrinkage rate of PET yarns in acetonitrile and in dimethylformamide. (*Source*: From Ribnick et al. [24] with permission of the publisher, Textile Research Institute.)

into the filament [24,35], under the assumption that shrinkage occurs only after complete penetration of the filament. The extent of polymer-solvent interaction can be deduced from the extent of the final or quasi-equilibrium shrinkage. As was pointed out above for density and swelling, the degree of shrinkage is also dependent on the solubility parameter of the solvent [23], which reflects the degree of polymer-solvent interaction. It should be mentioned, however, that equilibrium shrinkage is not always a reliable measure of the extent of interaction, since its value is influenced by the rate of molecular relaxation processes and the rate of crystallization or crystal growth interfering with these molecular relaxation processes. Depending on the nature of the solvent (its interaction and diffusion characteristics), the temperature, and other experimental variables, one or the other of the two processes—relaxation or crystallization—will be dominant in determining the ultimate shrinkage value. Solvent treatment under constant length conditions produces a shrinkage stress which under certain conditions passes through a maximum that reflects viscoelastic stress relaxation processes at longer times [24].

5.3 Dynamic Shrinkage

The observation that equilibrium shrinkage values obtained under isothermal conditions show a linear temperature dependence over a

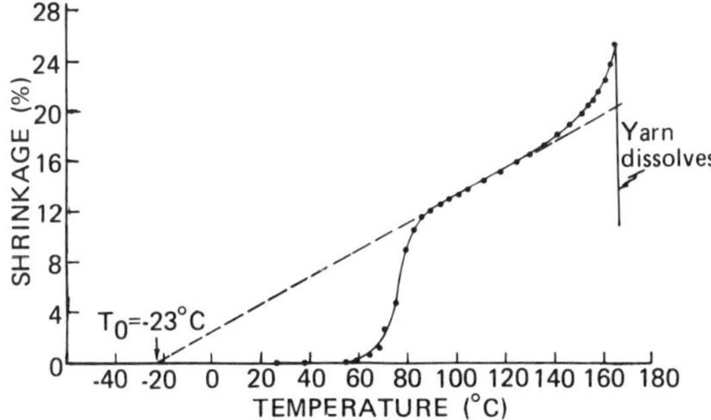

Figure 1.12 Dynamic shrinkage of a polyester yarn in γ-butyrolactone at a heating rate of 3.5°C per min and at a constant stress of 6.57 × 10^{-4} g/den. (*Source*: From Ribnick and Weigmann [25] by permission of the publisher, Textile Research Institute.)

significant temperature range resulted in a new approach, in which shrinkage is measured as a function of time at a programmed heating rate [25]. A dynamic shrinkage curve involving either thermal shrinkage or solvent induced shrinkage has essentially three regions, as illustrated in Fig. 1.12. The first region is a nonequilibrium region in which diffusion and shrinkage rates are substantially slower than the rate of heating. This is followed by a linear portion in which shrinkage and diffusion are much faster than the rate of heating,

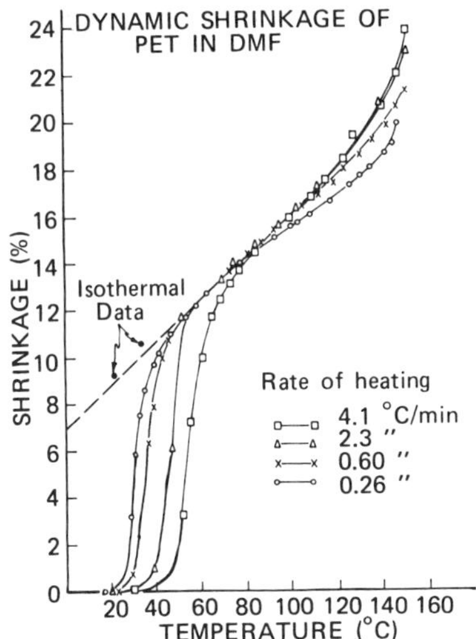

Figure 1.13 Dynamic shrinkage of polyester yarns in DMF at various rates of heating under a constant stress of 6.57 × 10^{-4} g/den. (*Source*: From Ribnick and Weigmann [25].)

resulting in equilibrium levels of shrinkage as the temperature is raised. The enhanced shrinkage rate in the third region, which occasionally results in dissolution of the fiber, is attributed to solvent melting of existing small crystallites and possibly a certain amount of recrystallization.

Extrapolation of the linear region of a dynamic shrinkage curve to a zero shrinkage level yields a temperature T_0 which appears to be an estimate of the effective glass transition temperature of the polymer-solvent system. While the extrapolated value depends to some extent on the rate of temperature increase, it has been shown that at heating rates of about 3-5°C per min, the extrapolated value agrees well with values obtained from isothermal equilibrium shrinkage values (Fig. 1.13). It should be pointed out, however, that this method of obtaining a temperature characteristic of the extent of polymer-solvent interaction is applicable only to a limited number of polymer-solvent systems. It is obviously important that the movement of solvated chain segments occur without interference from simultaneous crystallization and that the length of the linear region, which depends on the boiling point of

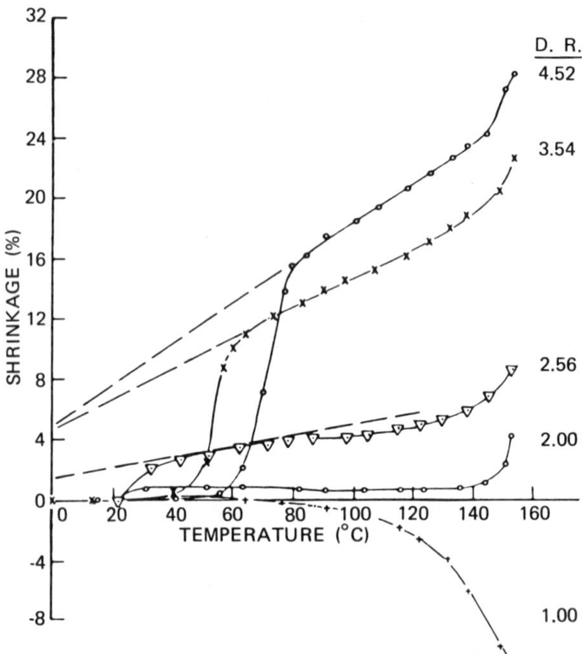

Figure 1.14 Dynamic shrinkage of polyester yarns of various draw ratios in DMF (heating rate 4.3°C per min, constant stress 4.5×10^{-4} g/den. (*Source*: From Rebenfeld et al. [1], by courtesy of Marcel Dekker, Inc.)

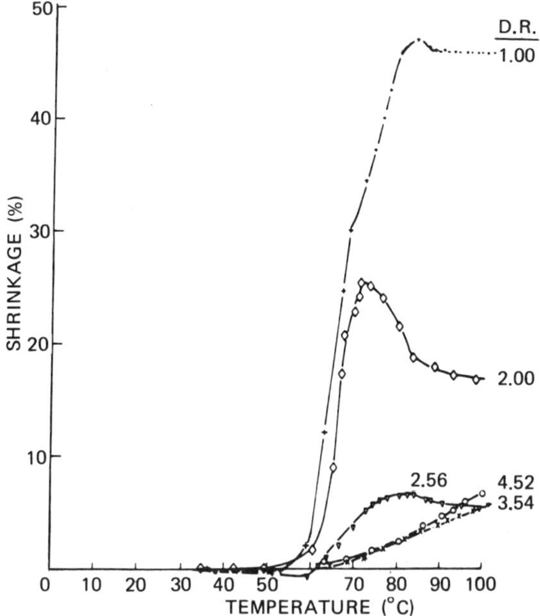

Figure 1.15 Dynamic shrinkage of polyester yarns of various draw ratios in water (heating rate 0.5°C per min, constant stress 4.5×10^{-4} g/den). (*Source*: From Rebenfeld et al. [1].)

the solvent and the nature of the polymer, be long enough to permit extrapolation to a reasonably accurate value.

Draw ratio affects the level of shrinkage of polyester in various solvents but does not affect the temperature dependence of equilibrium shrinkage levels or of the linear region in the dynamic shrinkage curve (Fig. 1.14). In other words, at least at draw ratios above 2, the extrapolated zero shrinkage temperature is essentially independent of the draw ratio. In noncrystalline PET, i.e., in fibers either undrawn or with a low draw ratio, penetration of an interactive solvent such as dimethylformamide (DMF) causes crystallization in the periphery of the filament as pointed out above and thereby rigidifies the filament before complete penetration and shrinkage can occur. In poorly interactive solvents, on the other hand, the effect of draw ratio is reversed, as shown in Fig. 1.15 for the dynamic shrinkage behavior of polyester filaments of various draw ratios in water. This difference in behavior has been explained in terms of the balance between molecular relaxation, both thermal and solvent induced, and the ability of the solvent to induce other structural alterations, such as SINC.

In an attempt to quantify the interactions between solvent and polymer, Weigmann and co-workers [24] have used the zero shrinkage

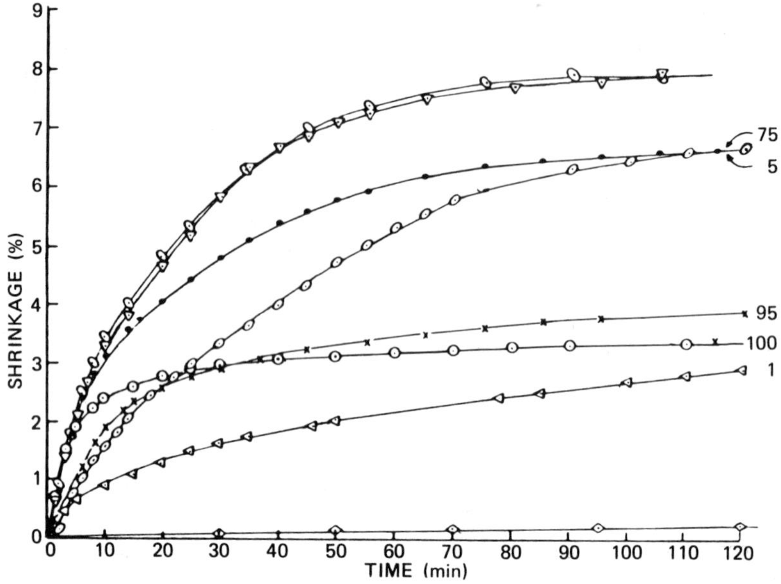

Figure 1.16 Isothermal shrinkage of polyester yarns in perchloroethylene-methanol mixtures (50°C). (*Source*: From Rebenfeld et al. [1].)

temperature obtained from the extrapolation of the linear region of dynamic shrinkage curves to establish the shift in the glass transition temperature, $\Delta T_g = T_g - T_0$, where T_0 is the glass transition temperature of the dry polymer. When ΔT_g is plotted against the solubility parameter of the solvent, a bimodal distribution such as those already discussed is approximated.

5.4 Solvent Mixtures

The effects of mixtures of organic solvents on the shrinkage behavior of polymeric fibers depend on the extent to which the components interact with each other. Such intersolvent interactions can lead to appreciable synergism, particularly when one of the components is characterized by strong intermolecular bonding among its own species, such as occurs in water or methanol. The effect has been demonstrated by Weigmann and others [82] in the isothermal shrinkage behavior of PET in mixtures of perchloroethylene and methanol, shown in Fig. 1.16. It is clear that mixtures with intermediate levels of composition have a significantly larger degree of interaction with PET than either one of the two pure components. In contrast, the shrinkage behavior of PET in mixtures of noninteracting solvents, such as perchloroethylene and trichlorobenzene, is directly proportional to the composition of the solvent mixture. The application of the solubility parameter principle to the analysis of the shrinkage of polyester filaments in solvent mixtures has been attempted by Knox [83]. It has been difficult, however, to apply this approach satisfactorily, especially where significant interaction between the components of the solvent mixture occur.

6. SOLVENT INDUCED CHANGES IN THE PHYSICAL PROPERTIES OF FIBERS

6.1 Fiber Structure-Property Relationships

An understanding of the effects of solvents on the mechanical properties of polymeric fibers requires a structural model that quantitatively describes relationships between polymer structure and mechanical properties. To a first approximation, the semicrystalline polymer as it exists in polymeric fibers has been considered a complex, two-phase system in which crystalline domains are dispersed in an amorphous matrix. The general applicability of this type of model has been discussed in considerable detail, and alternative models have been suggested that describe mechanical behavior in a more satisfactory manner.

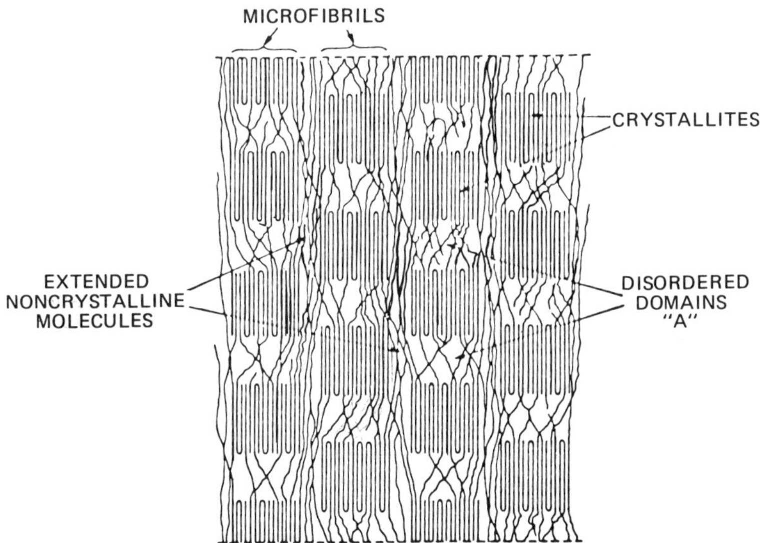

Figure 1.17 Schematic structure of PET fibers (fiber axis vertical). (*Source*: From Prevorsek et al. [85], by courtesy of the publisher, Marcel Dekker, Inc.)

Prevorsek [84,85] has postulated that oriented noncrystalline domains consisting of more or less extended polymer chains exist between microfibrils (Fig. 1.17). The proportion of the fiber structure made up of such oriented interfibrillar domains depends on the draw ratio of the fiber. These extended amorphous domains are similar to the extended or interlamellar tie molecules postulated by Peterlin [86, 87] and may be considered the essential elements of fiber structure with regard to strength and deformation processes. It is these domains that are plasticized and restructured as a result of polymer-solvent interactions and therefore have a major effect on the properties of the solvent treated fibers, both in the presence of the solvent and after its removal.

Depending on the extent of polymer-solvent interaction and on the prior thermomechanical history of the fiber, a large range of solvent induced structural changes in the polymeric fibers can occur, as has been discussed to some extent in previous sections. The effects of plasticization and of solvent induced structural changes on the mechanical properties of polymeric fibers are discussed in the following paragraphs.

6.2 Plasticization of Polymeric Fibers by Solvents

It is assumed that in most cases solvents do not penetrate the crystalline domains in polymeric fibers and therefore do not affect the mechanical properties of these domains. This implies that changes in the mechanical properties of semicrystalline polymers due to swelling solvents are brought about by plasticization of the amorphous domains [88]. Penetration of the solvent into the amorphous regions involves the breakdown of intermolecular bonds and produces increased segmental mobility of the polymer chains, which is reflected in a lowering of the glass transition temperature. During the deformation of such plasticized polymeric fibers, the forces that are transmitted through the swollen amorphous layers to the crystalline domains are small and do not cause large-scale deformation of the crystallites. Deformation in a plasticized polymeric fiber, therefore, mainly involves sliding, rotation, and possibly physical separation of crystalline lamellae [86].

Similar increases in the segmental mobility of the amorphous domains in the dry or unplasticized state can be obtained by increasing the temperature. However, the deformation behavior of semicrystalline polymeric fibers in the presence of a swelling agent and its deformation at a higher temperature in the absence of a swelling solvent are significantly different, as pointed out by Peterlin [86]. The crystal lattice is essentially unaffected by the solvent but considerably weakened by higher temperature. Thus, drawn polymeric fibers have significantly different properties depending on whether they have been drawn in the presence of a solvent or drawn at elevated temperature.

If a model like that of Prevorsek adequately describes polymeric fibers, increased interaction between solvent and polymer should lead to more extensive penetration of the oriented amorphous domains. As was pointed out in previous sections, these domains are believed to control the contraction behavior of the oriented polymer, and increasing levels of shrinkage are observed as higher levels of orientation are affected. This can be brought about either by using solvents that are more interactive or by increasing the temperature at which the solvent is permitted to interact with the polymeric fiber [25].

The mechanical properties of solvent-swollen fiber structures have been investigated by Weigmann and co-workers [24, 26], especially for the case of highly drawn, semicrystalline polyester filaments. A general correlation was found between the ability of the solvent to shrink the fiber and the initial modulus and yield stress in the solvent-swollen filaments. Both Young's modulus and, particularly, the yield stress are closely related to molecular mobility in the amorphous domains, as discussed by Hookway [89] and by Bunn and Alcock [90]. Fig. 1.18 shows load-extension curves for polyester yarns after swelling at room temperature in a number of interacting solvents. These data clearly illustrate variations in the plasticizing effects of the different solvents [24]. It is typical that the force at break is relatively unaffected by the solvent treatments, while the extensibility varies significantly. It appears that the disorientation of the oriented extended noncrystalline domains, which is associated with the shrinkage of the filament in the solvent, is easily reversible on extension, while the regions that have not been penetrated by the solvent maintain their original properties. This is illustrated in Fig. 1.19 for polyester fibers swollen in methylene chloride and tetrachloroethane. Treatment

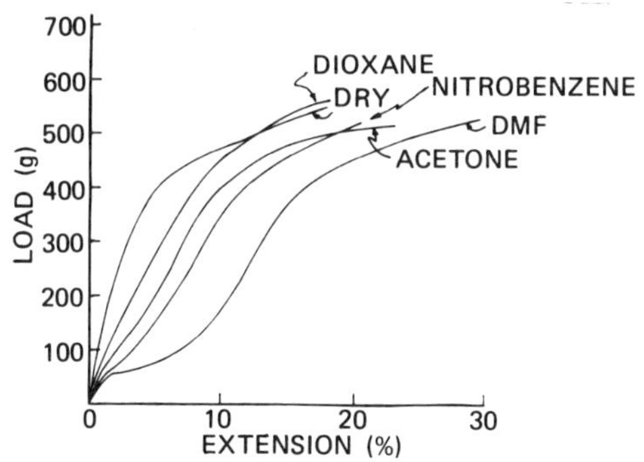

Figure 1.18 Load-extension curves of polyester yarns in dioxane, nitrobenzene, DMF, and acetone at 25°C. (*Source*: From Ribnick et al. [24] with permission of the publisher, Textile Research Institute.)

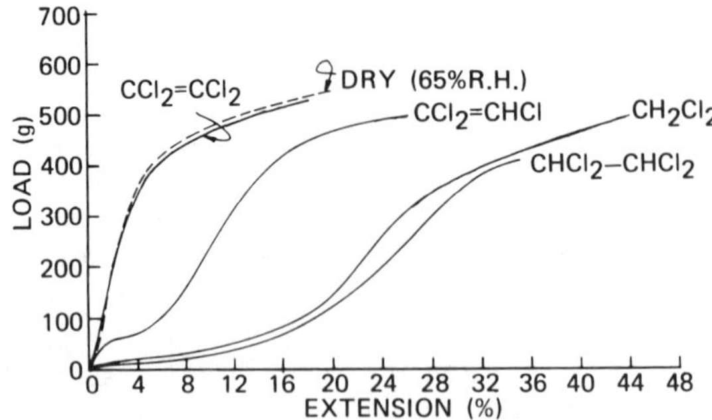

Figure 1.19 Load-extension curves of polyester yarns in several chlorinated solvents at 21°C. (*Source*: From Ribnick et al. [24].)

in both these solvents results in large-scale penetration and plasticization of the fiber, leading to highly swollen structures which display essentially rubbery behavior in their initial deformation range, with extremely low values of Young's modulus and complete disappearance of the yield stress. Removal of the solvent from the polymer results in the partial restoration of the original properties, depending on the extent of fiber shrinkage that has taken place. The removal of the solvent by drying, however, is not always complete, and the plasticizing by residual solvent contributes to incomplete recovery of mechanical properties.

6.3 Effects of Solvents on Fiber Mechanical Properties

As pointed out above, the mechanical properties of solvent treated fibers are reflections of the structural modifications that the solvent treatment has induced. These modifications depend on the nature of the polymer-solvent interaction, the treatment time, and the treatment temperature. Weigmann and co-workers [27] have investigated the effects of treatments with dimethylformamide (DMF) as a typical strongly interacting solvent for polyester fibers. DMF causes substantial swelling of the polyester structure, resulting in relaxation of built-in stresses and fiber shrinkage. As the treatment temperature increases, crystallization and melting/recrystallization of imperfect crystalline domains begin to occur, and the swollen structure is somewhat stabilized by the newly formed crystallites. Upon removal of the solvent the structure does not completely collapse: microvoids and voids are formed, and a structure of considerable microporosity is produced which is clearly discernible in small-angle x-ray scattering patterns. This is in sharp contrast with the effects of purely thermal treatments, which also result in relaxation of built-in strains and melting/recrystallization, but do not produce the extensive void formation.

The initial disorientation in the structure which is not accompanied by crystallization is reflected in linear decreases in initial modulus and yield stress with a reciprocal shrinkage function, as shown in Fig. 1.20. At higher shrinkage values a discontinuity in this linear relationship is observed, indicating that a more extensive reorganization of the structure occurs at higher treatment temperatures, involving significant recrystallization. It is interesting to note that this discontinuity occurs at the same shrinkage value for both heat and DMF treatments.

Initial shrinkage involving only a minor disorientation of the extended oriented polymer chains can be recovered in subsequent extension. However, the extent of recovery depends on the structural changes that are produced by the treatment and is therefore a reflection of the extent of crystallization, which usually increases at higher shrinkage levels and higher treatment temperatures. This aspect has been discussed by Ueda and Nukushina [91] and by Prevorsek et al. [85]. Shrinkage produced by solvent treatments is much more easily recovered in subsequent extension than shrinkage after thermal treatments. This high reversibility of solvent induced shrinkage is a reflection of the nature of recrystallization occurring under the influence of solvents. It has in general been observed [92, 93] that solvent treatments produce smaller crystallites, and that the structure retains significant mobility due to the fact that this recrystallization occurs in a more or less swollen state. The partial collapse of the swollen structure upon removal of the solvent eliminates a significant part of the strains produced during swelling and crystallization. In general,

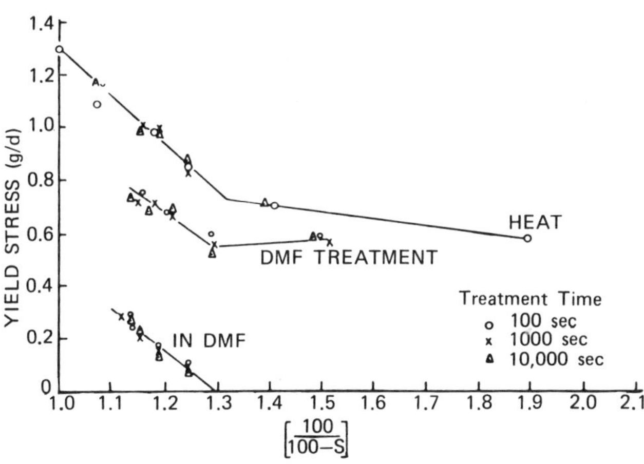

Figure 1.20 Yield stress (based on linear density of shrunk yarn) of DMF and heat treated polyester yarns as a function of shrinkage (S) related to changes in linear density [100/(100 - S)]. (*Source*: From Rebenfeld et al. [1].)

these observations are consistent with suggestions by Wilson [76] that the basic thermal shrinkage mechanism for a drawn fiber is a "rapid, rubberlike contraction associated with disorientation in the amorphous phase. That disorientation will be followed by a crystallization stage during which the formation of IR detectable folded chain structures will be favored only if sufficient time is allowed." During thermal treatment rapid and extensive crystallization occurs, but swelling and subsequent solvent induced crystallization seem to require significantly more time. The mechanical properties of solvent treated polyester filaments reflect this behavior in their low moduli and high extensibilities.

Treatments under constant length conditions have been used to impart dimensional stability to structures made from artificial fibers. These heat setting procedures (i.e., relaxation under constant length conditions) stabilize the structure to subsequent treatments under free shrinkage conditions. The possibility of using solvent treatments under constant length conditions to set filaments or fabrics has been investigated by Weigmann et al. [94] and by Venkatesh et al. [95]. Although increasing stabilization is observed with increasing treatment temperature in the solvent, significant residual shrinkage under thermal conditions or in boiling water suggests that solvent setting is only partly successful. Weigmann and co-workers [94] suggest that the incomplete stabilization is a reflection of the smaller number of stabilizing crystals that are formed during solvent treatments under constant length conditions than under thermal setting conditions, as well as of the wider distribution of crystallite size. The fact that stabilization occurs in the swollen state and that there is residual mobility upon removal of the solvent also contributes to incomplete stabilization.

It is interesting to note that structural modifications introduced by a solvent treatment at elevated temperatures can be overcome by subsequent thermal treatments of sufficient severity. It has been suggested that solvent treatments under free shrinkage conditions lead to a disoriented fiber structure which is stabilized by crystallites formed during the solvent treatment and that this stabilization results in cavitation and void formation upon subsequent removal of the swelling medium [27]. Differential scanning calorimetry (DSC) and other techniques have been used to establish the limits of thermal stability of the structure produced by the solvent treatment [94,96]. The stabilizing crystallites are reflected in a premelting endotherm in the DSC curves at approximately 70°C above the treatment temperature. This characteristic, diffuse, premelting endotherm observed after solvent treatments suggests the formation of crystallites with a relatively wide distribution of thermal stabilities which, as pointed out above, only partially stabilize the microporous fiber structure. Heat treatments at temperatures exceeding the temperature of the premelting endotherm destroy the stabilizing crystallites of the microporous structure and lead to collapse of the voids. The effects that the formation of voids and their subsequent collapse due to heat treatments have on the dyeability of polyester yarns will be discussed in a subsequent section.

Prior thermal treatments of polyester fibers can stabilize the structure to subsequent solvent modification [97]. Depending on the extent of prior heat treatments, the ability of the fiber to swell and recrystallize under the influence of the solvent is decreased, and after heating at sufficiently high temperatures the solvent treatments are no longer capable of modifying the structure at all [94]. The extent to which heat treatments stabilize is reflected in the sharp characteristic premelting endotherm slightly above the treatment temperature that is produced in the diffential scanning calorimetric curve. Solvent treatments cannot overcome the effect of prior heat treatments as long as the stability of the stabilizing crystallites, which is reflected in the temperature of the premelting endothermal peak, can withstand the combined chemical and thermal energy provided by the solvent treatment.

6.4 Effects of Solvents on Dynamic Mechanical Properties of Fibers

The dynamic mechanical properties of a polymeric fiber are reflected in the response of the fiber to periodic low-level deformations. In the dynamic mode of deformation, energy is either stored or dissipated in the polymer, and one obtains measurements for elastic behavior and for mechanical damping. The process of energy dissipation is expressed by the loss modulus and is a measure of the segmental mobility in the polymer structure. A number of relaxation maxima appear when the dynamic mechanical behavior is measured over an extended temperature range. The so-called α-relaxation peak, occurring at the highest temperature, reflects the large-scale motion of chain segments within the amorphous domains of the polymer. This peak has been associated with the glass transition temperature.

The torsion pendulum [98] and the Rheovibron direct-reading viscoelastometer [99] are generally used to determine dynamic mechanical properties of fibers at low and high frequencies, respectively. Disruption of intermolecular bonding and higher segment mobility in solvent-plasticized polymeric fibers leads to a downward shift in the temperature of the α-relaxation. These effects of solvents are well known and have been reported by a number of workers [100-102]. Dimarzio and Gibbs [103] have given a molecular interpretation of the lowering of the α-relaxation temperature by interactive solvents.

In polyamides, a new relaxation process has been observed by Illers [104] and Goldbach [105], which is associated with the presence of water. The appearance of such new peaks in the relaxation spectrum of polyamides has also been observed with other solvents, such as methylene chloride and acetic acid esters. It appears to be associated with the formation of strong polar bonds between the solvent and appropriate groups in the polymer.

Measurement of the dynamic mechanical properties of polymer films or fibers in contact with liquid solvents requires a modification of the Rheovibron instrument. This has been successfully achieved by a number of different investigators [106-109]. Apparently, the fact that the polymeric substrate is immersed in a liquid medium has no effect on the dynamic mechanical properties except for very viscous liquids at very high frequencies [110], so that it is possible to establish the effects of solvent penetration on the dynamic properties of the fiber. These measurements of dynamic mechanical properties in the presence of solvents are of considerable interest for various textile processes, especially dyeing, where the effect of nonaqueous solvents and dyeing assistants on the segmental mobility in the fibers is critical for diffusion and uptake of dyestuffs and finishing agents. Fujita et al. [111] have shown that "at temperatures above the α-relaxation temperature, correlations exist between the diffusion coefficient of penetrants, and the steady state viscosity of the polymeric substrates." On this basis, quantitative correlations have been established between the diffusion coefficients of dyes in polymers and the dynamic loss modulus. Dumbleton and co-workers [112] have reported such a correlation for a series of polyester fibers of various draw ratios and certain disperse dyes:

$$\ln(D_T/RT) = -B \ln(E''/\omega) + [C - B \ln(1/3)] \tag{18}$$

where ω is the frequency at which the dynamic loss modulus E" is measured, and B and C are constants. This equation has also been successfully applied to the diffusion of acid dyes in drawn nylon 66 fibers [113] as well as to the dyeing of polyester fibers from organic solvents [109].

7. SOLVENTS IN DYEING AND FINISHING OPERATIONS

The conventional dyeing of most textile substrates is an energy-intensive operation requiring the heating of significant dye bath volumes for extended periods of time to attain sufficient penetration of the dyestuff into the fiber structure. This is especially true for the dyeing of polyester fibers or polyaramid fibers (Nomex, Kevlar), which are particularly difficult to dye because of their high glass transition temperatures. A number of approaches have been used to utilize fiber-solvent interactions to decrease the energy requirements of dyeing, and a vast published literature exists which cannot be covered in this brief review. An attempt will be made to illustrate with appropriate examples the various methods in which polymer-solvent interactions have been utilized for this purpose.

There are essentially two ways in which solvents can be used to improve the transport of dyestuff molecules into the fiber structure. First, the solvent can modify the fiber structure in a more or less temporary way by directly participating in the dyeing process either as an addition to the dye bath or as the major dyeing medium. Second, solvents can be used to modify the fiber structure irreversibly in a treatment prior to the dyeing process.

7.1 Solvent Dyeing

The use of organic solvents as dye bath media became quite a popular concept during the past decade, when it was realized that fresh water availability and wastewater disposal would be significant and expensive problems for the textile industry. Dyeing from organic solvents could also represent considerable energy savings, since solvents usually have a much lower heat of evaporation than water and can be more or less completely recycled by means of standard recovery operations, thus eliminating costly drying after aqueous dyeing as well as potential water pollution problems. One of the disadvantages of solvent dyeing and finishing, on the other hand, is possible retention of the solvents by the textile substrate and their subsequent slow release, which can lead to air pollution and toxicity problems that have to be taken into consideration. Hofstetter [114] reviewed the literature on solvent dyeing through 1970 and summarized the current situation in a comprehensive review article. A more recent summary of solvent textile processing has been given by Skelly [115].

The dyeing of polyester substrates from solvents using dyestuffs that are conventionally used as disperse dyes in aqueous dyeing has received particularly close attention. Gebert [116] has examined the dyeing behavior of polyester with various disperse dyes dissolved in perchloroethylene, giving special attention to the effects of added auxiliary solvents on the color yield and the distribution of the dye between the liquid and solid phases. Exhaust dyeing of polyester from organic solvents has been critically reviewed by Kothe [117], who points out that the considerable problems with exhaust dyeing of disperse dyes from perchloroethylene are mainly associated with the unfavorable partition coefficient.

The underlying principle of exhaust dyeing is the partition of dyestuff between the fiber and the dyeing medium. At equilibrium this partition can be expressed according to Nernst's partition law:

$$K = \frac{C_f}{C_s} \tag{19}$$

It is obvious that a high distribution coefficient K is required for exhaust dyeing; in other words, the concentration of the dye in the fiber C_f should be much greater than the solubility of the dye in the dyeing medium C_s. The exhaustion of a dye bath is given by the equation:

$$E = \frac{K}{K + L} \qquad K = \frac{EL}{1 - E} \tag{20}$$

The smaller the liquor ratio L, the higher will be the degree of exhaustion E at a given value of K. This was illustrated by Kothe [117], who showed (Fig. 1.21) that an acceptable exhaustion of about 90% can be achieved only at high K values or at very low liquor ratios. Perchloroethylene is for a variety of reasons still considered the most suitable organic solvent for dyeing purposes, but conventional disperse dyes have low K values in perchloroethylene and correspondingly unsatisfactory bath exhaustions. Although the addition of small amounts of water improves the degree of exhaustion significantly, disperse dyeing of PET from solvents is currently no alternative to aqueous dyeing methods [117]. Successful solvent dyeing will require the synthesis of new dye systems having low solubility in organic solvents and good affinity for the fiber, so as to provide adequate distribution coefficients for good dye exhaustion levels.

Harris and Guion [118,119] have discussed the application of the solubility parameter concept for the proper selection of solvents, dyes, and polymers in designing workable solvent dyeing systems. Fig. 1.22 is a schematic representation of the dilemma that the dyer faces. He or she must choose advantageous parameters either kinetic or equilibrium. Disperse dyes usually have solubility parameters close to that of the polymer, as shown by Ibe [120] for the case of nonionic dyes

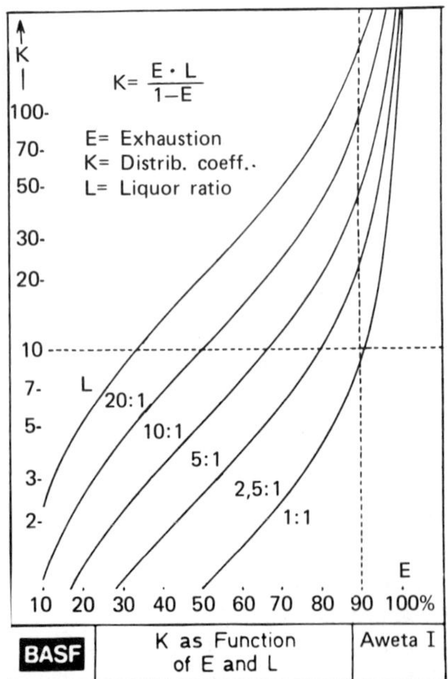

Figure 1.21 Distribution coefficients of dyestuff between dye bath and fiber as a function of exhaustion level for various liquor ratios. (*Source*: From Kothe [117] by permission of the publisher, The American Association of Textile Chemists and Colorists.)

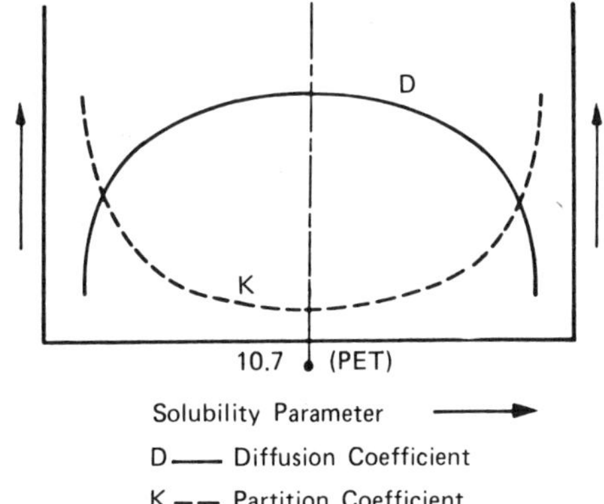

Figure 1.22 Schematic representation of diffusion and partition coefficients as a function of solubility parameter of solvent. (*Source*: From Harris and Guion [118] by permission of the publisher, The American Association of Textile Chemists and Colorists.)

used for the dyeing of cellulose acetates. It can be assumed, therefore, that disperse dyes with high distribution coefficients from an aqueous environment have solubility parameters close to that of polyester. Solvents chosen as dyeing media must interact sufficiently with the polymer to cause swelling of the fiber and thereby increase the diffusion coefficient of the dyestuff. This level of interaction between solvent and polymer can be obtained only if they have similar solubility parameters. At the same time, however, the solubility of a dye in a solvent having a similar solubility parameter will also be high, resulting in a very low distribution coefficient between the solvent phase and the polymer. This is indeed the case for the dyeing of polyester with disperse dyes from a perchloroethylene environment, even though perchloroethylene has only a relatively limited ability to interact with polyester [26]. Even lower distribution coefficients result with solvents, such as methylene chloride or dimethylformamide, that interact strongly with polyester and are actually used to extract dyestuffs from the fibers. If a solvent is chosen with a solubility parameter significantly different from that of the polymer and the dyestuff, the distribution coefficient would be expected to be high, i.e., in favor of deposition of dyestuff inside the fiber. In this case, however, interactions of the solvent with the polymer are low, resulting in only a minor decrease in the glass transition temperature, a low level of swelling, and a low diffusion coefficient. The typical example is the dyeing of polyester fibers with disperse dyes from an aqueous medium.

A number of possibilities exist that may provide a way out of this apparent impasse inherent in solvent dyeing. Obvious choices would be to dye at an extremely low liquor ratio or to use vapor-phase dyeing, which will be discussed in Sec. 7.2. Another alternative would be the development of new dyes with high distribution coefficients in a suitable solvent, i.e., a solvent with sufficient interaction with the polymer to permit acceptable diffusion coefficients. As Harris and Guion have pointed out [118,119], this situation is inconsistent with solubility parameter theory. The only hope that solubility parameter theory offers in this regard is an appropriate choice of the three-component system of polymer, solvent, and dyestuff with regard to the three composite solubility parameters for hydrogen bonding, polar, and dispersion forces of interaction. First steps in this direction have been made in a recent publication by Siddiqui [121]. Finally, there is the possibility of using cosolvents such that the solubility of dyestuffs in the mixture is low but the mixture or one of its components interacts with the polymer to a suitable degree. This principle is employed in the so-called STX dyeing system [122], which is used for the dyeing of nylon fabrics. In this process, methanol is used as a cosolvent with the major dye bath component, perchloroethylene, in order to solubilize the dyestuff in the dye bath and at the same time plasticize the polymer structure. During the dyeing process the cosolvent is slowly evaporated, thereby continuously increasing the distribution coefficient of the dyestuff, which is essentially insoluble in the major solvent component, perchloroethylene.

7.2 Solvent Pretreatments

The pretreatment of polymeric fibers with various solvent systems can lead to fiber structures which are presensitized to subsequent dyeing processes. The improved dyeability of these pretreated fibers is achieved either through the retention of solvent in the fiber structure, which then acts like a carrier in subsequent dyeing processes, or through a permanent modification of the fiber structure.

Solvent Retention

It is well known that, especially in the case of chlorinated hydrocarbons, significant amounts of solvent are retained by the fibers after solvent treatments at elevated temperature; the solvent is released rather slowly [123-128]. Rieker and Terlinden [127] have studied the uptake of various chlorinated hydrocarbons by polyester yarns and the resulting effects on dyeability. As shown in Fig. 1.23, significant solvent uptake does not start until solvent treatment temperatures of about 50°C have been exceeded. Rieker and Terlinden attribute the onset of solvent uptake to a lowering of the glass transition temperature of PET by perchloroethylene to a temperature of about 50°C. At a dyeing temperature of 120°C perchloroethylene content is seen to have relatively little effect on dye uptake relative to that of the untreated

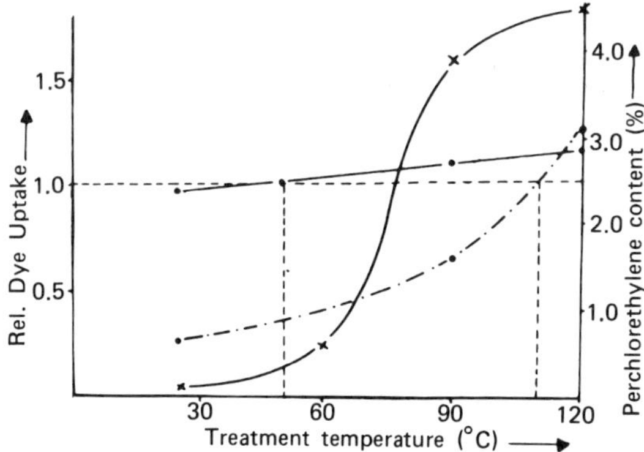

Figure 1.23 Effect of perchloroethylene treatment temperature on:
— X — Perchloroethylene content of pretreated polyester
— · — Dye uptake of pretreated polyester at 102°C relative to uptake of untreated sample
-·-·-·-·- Dye uptake of pretreated polyester at 98°C relative to uptake of untreated sample with carrier in dyebath
(*Source*: From Rieker and Terlinden [127] with permission of the publisher, *Melliand Textilberichte*.)

sample. For a dyeing temperature of 98°C, comparisons are made with carrier assisted dyeing of the untreated sample, since little dye is taken up at this temperature without a carrier. The perchloroethylene pretreated sample takes up increasingly more dyestuff as the perchloroethylene content increases, and the dye uptake level becomes comparable to that in carrier dyeing at a temperature of 110°C when the perchloroethylene content is about 4.5%. Recently, Bhattacharje [129] has shown that at high temperatures perchloroethylene is retained in significant amounts even after very short treatment times, producing improved dyeability. Bredereck and co-workers [126] reported that at 120°C equilibrium perchloroethylene sorption levels can indeed be reached within a few minutes. However, at these relatively high temperatures, relaxation and secondary crystallization can occur, which may substantially modify the ability of the fiber to absorb dyestuff. Thus, it is difficult to distinguish between the contributions to improved dyeability from retained solvent and those from structural modifications.

Smith and Barwick [130] have shown that very short, high-temperature treatment in 1,2,4-trichlorobenzene leads to the introduction of a significant amount of this carrier into the fiber structure, with substantial improvements in dyeability. This treatment procedure eliminates the need for carrier in the dye bath and can be carried out in a continuous manner.

Solvent as Transport Medium

The introduction of carrier via pretreatment either from an aqueous environment or from a solvent, usually involving elevated temperatures to permit diffusion of the carrier into the fiber structure, has been discussed by a number of workers. It has recently been shown [131, 132] that methylene chloride can transport carriers into the polyester fiber structure at low temperature. A brief, room-temperature pretreatment with biphenyl carrier in methylene chloride apparently deposits sufficient amounts of carrier at or near the fiber surface to produce excellent dye penetration in subsequent standard, aqueous disperse dyeing procedures under atmospheric conditions (Fig. 1.24). In this case it has been shown that the deposition of the carrier is the decisive element in the pretreatment and that methylene chloride, at least under these short-term conditions, has only a minor function in affecting subsequent dyeability. While this carrier pretreatment produces dyeing results that are similar, or maybe even superior in terms of dyestuff penetration, to dyeings carried out with the carrier dispersed in the dye bath, such improvement is not necessarily produced if carriers are deposited from an aqueous solution [133].

In general, it can be said that the retention of solvents that can interact with the polymer, or the introduction of carriers that will

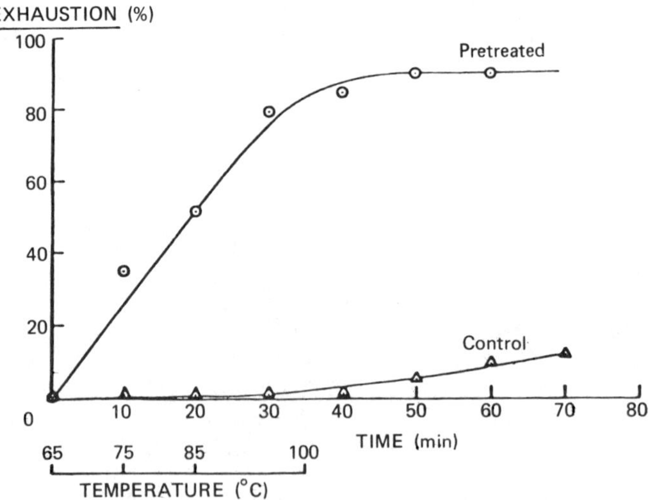

Figure 1.24 Dyebath exhaustion profiles for 1.5% (owf) Bucron Rubine 2BNS on Dacron 56 fabric pretreated with biphenyl-$MeCl_2$. (*Source*: From Matkowsky et al. [131] by permission of the publisher, The American Association of Textile Chemists and Colorists.)

fill essentially the same function, leads to a temporary lowering of the glass transition temperature of the polymer and thereby improves dye diffusion kinetics in subsequent conventional aqueous dyeing. An important aspect, however, is that these compounds have to have relatively low volatility and definitely low solubility in water in order to remain inside the fiber structure long enough to provide increased segmental mobility and dye diffusion rates during the dyeing process. Low-boiling solvents such as methylene chloride are most likely removed from the fiber structure during the early phase of the dyeing process and are consequently not available in sufficient quantity to produce the desired effect on the glass transition temperature during the dyeing process. The unique properties of methylene chloride in terms of rapid diffusion and extremely high interaction with polyester, on the other hand, have been recognized and utilized in a number of pretreatment or aftertreatment processes. The use of methylene chloride as a transport medium for dyestuffs has recently been reported in the patent literature [134-136]. In this case vapors of halogenated hydrocarbons, usually methylene chloride, are used to transport dyestuffs that had previously been deposited on the fiber surface into the interior of the fiber under very mild, low-temperature conditions.

Permanent Structural Changes

For solvent pretreatments of textile fibers to be of interest to the dyer, they have to result in faster dyeing, deeper shades, and better fastness properties than can be achieved using conventional dyeing processes. These pretreatments would have to be carried out under conditions that are technologically and environmentally acceptable, and they would have to produce a cost advantage over current dyeing procedures. Furthermore, fiber modifications would have to be achieved without downgrading other important fiber properties, such as mechanical, aesthetic, or other end-use characteristics.

As was pointed out in previous sections, substantial changes in fiber structure can be induced during treatment with organic solvents, leaving a structure that has a permanently lowered glass transition temperature and a much more open polymer structure, resulting in improved dyeability. In the case of polyester fibers, Weigmann and co-workers [27] have shown that treatment with an interactive solvent such as dimethylformamide causes considerable swelling of the fiber and recrystallization in the swollen state. The polymer crystallites produced in the swollen state tend to stabilize the polymer structure in this condition and prevent its complete collapse upon subsequent solvent removal, leaving microvoids in the fiber structure which are detected as diffuse, small-angle x-ray scattering (Fig. 1.25). This more open and void-containing fiber structure not only permits increased rates of dye transport, but also enhances saturation dye uptake. The large increase in dye uptake observed in Fig. 1.26 occurs

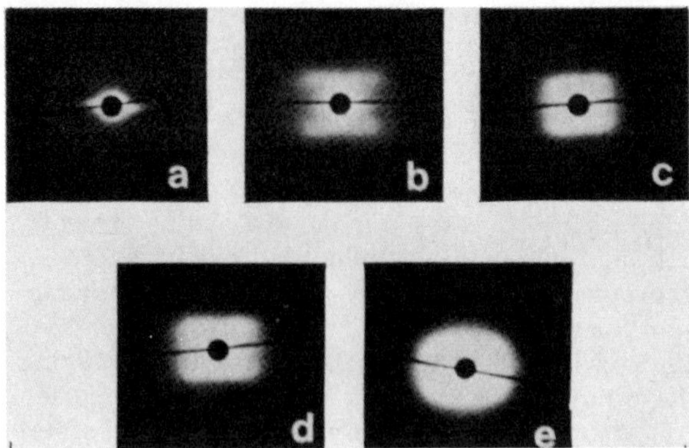

Figure 1.25 Small-angle x-ray patterns of polyester yarns after DMF treatments under unrestrained conditions for 120 sec followed by rinsing in water at 100°C. (a) untreated, (b) 120°C, (c) 130°C, (d) 135°C, (e) 140°C. (*Source*: From Weigmann et al. [137] by permission of the publisher, Textile Research Institute.)

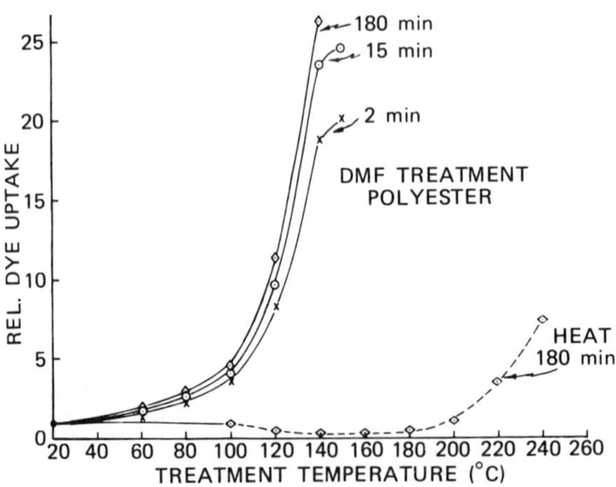

Figure 1.26 Relative uptake of Resolin Red BBL by DMF treated polyester yarns (unrestrained) as a function of treatment temperature. (*Source*: From Weigmann et al. [27] with permission of the publisher, Textile Research Institute.)

at treatment temperatures at which major void formation is observed. While there appears to be a lowering of the glass transition temperature as a result of these treatments, the major effect on dyeing behavior is probably the large increase in internal surface which permits increased deposition and possibly aggregation of dyestuff molecules. The porous structure produced by the solvent treatment has limited thermal stability, so that thermal aftertreatments [94-96] can be used to collapse the structure after the dyestuff has been introduced, thus effectively trapping the dyestuff molecules and providing good fastness properties.

A similar improvement in dyeability can be achieved in aromatic polyamides such as Nomex, which is well known for its problematic dyeing behavior [138,139]. Treatment with highly polar solvents such as dimethylsulfoxide, dimethylacetamide, or dimethylformamide produces a very open, porous structure with considerable void content as reflected in small-angle x-ray scattering. These solvent treated aramid fibers can be dyed rapidly in aqueous dye baths under atmospheric conditions without any carrier, and, depending on the severity of the solvent pretreatment, the fiber cross sections are completely penetrated under rather mild dyeing conditions.

It should be pointed out that the opening and disorientation of the polymer structure required for improved dyeability is in all cases accompanied by a decrease in mechanical properties, and an optimization of the pretreatment has to be achieved for the required dyeability-strength balance. Void-containing structures are an extreme manifestation of more open or porous structures resulting from treatment of semicrystalline polymeric fibers in highly interactive solvents. Pretreatment of nylon 6 and 66 filaments with interactive systems, such as aqueous and nonaqueous phenol solutions, results in significant structural changes and enhanced dye transport rate as well as enhanced saturation dye uptake [140]. It has been shown that the increased rate of dye diffusion in the treated nylon yarns can be attributed to greater polymer segmental mobility, as reflected in decreased dyeing transition temperatures. Without producing microvoids, the treatment nevertheless leads to a more open structure, as reflected in decreased orientation in the noncrystalline domains.

7.3 Solvent Scouring and Finishing Processes

While dyeing from a solvent medium has not been very widely accepted in the industry, such is not the case for fabric finishing from organic solvents. A significant number of processes are already in operation. Finishing processes from organic solvents can generally be assigned to one of three categories [141]:

1. Extraction, scouring
2. Applications to the fiber surface
3. Penetration of finishing medium into the fiber cross section

Organic solvents are particularly suitable and successful in extraction and scouring, and a number of solvent-based scouring and related processes are now in use throughout the textile industry. A typical example is the ICI Markal process, which permits scouring, simultaneous scouring and desizing, or scouring, desizing, and partial bleaching in one rapid, open-width step. This type of process appears to be competitive with current aqueous preparation methods and in addition gives excellent results in terms of uniformity, reproducibility, and high absorbency [142]. Chlorinated solvents, long used in the dry-cleaning industry, are also used in these processes because of their wetting properties, their good solvent power, and their chemical inertness.

Deposition of various finishes onto the fiber surface from solvents is a widely accepted practice, and materials such as stain repellents, hand modifiers and softeners, antistatic finishes, etc., are frequently deposited from organic solvents. A review of the literature up to 1970 is given by Hofstetter [114]; a more recent summary of current technology can be found in Reinert [143].

CONCLUSION

It is doubtful that we will see a major shift in the industry to solvent-based operations in the foreseeable future. Not enough is known about the ecological and toxicological properties of the chlorinated hydrocarbons that are being used, and recent trends in regulatory efforts in the United States do not seem to be encouraging in this respect. However, as more information becomes available and as a better understanding of the potential for solvent-based dyeing and finishing operations is developed, it appears likely that the use of organic solvents in certain areas of the textile industry can contribute significantly to the solution of ecological, economic, and technological problems.

ACKNOWLEDGMENTS

The author gratefully acknowledges advice and editorial help from Dr. Harriet Heilweil and Mrs. Nicolette Schwartzman.

REFERENCES

1. L. Rebenfeld, P. J. Makarewicz, H.-D. Weigmann, and G. L. Wilkes, *J. Macromol. Sci.—Rev. Macromol. Chem. C15*, 279 (1975).
2. R. Hagege, *Bull. Scient. ITF 6*, 77, 203 (1977).
3. H. Burrell, *Off. Dig. Fed. Soc. Paint Technol. 27*, 726 (1955).
4. S. A. Chen, *J. Appl. Polym. Sci. 15*, 1247 (1971).
5. J. L. Gardon, *J. Paint Technol. 38*, 43 (1966).

6. J. H. Hildebrand and R. L. Scott, *Solubility of Nonelectrolytes*, Reinhold, New York (1950).
7. G. Scatchard, *Chem. Rev. 8*, 321 (1931).
8. J. L. Gardon, *Encyclopedia of Polymer Science and Technology*, Vol. 3, 833 (1965).
9. I. Prigogine, *The Molecular Theory of Solutions*, North-Holland, Amsterdam (1957).
10. C. M. Hansen, *J. Paint Technol. 39*, 104 (1967); *Ind. Eng. Chem. Prod. Res. Dev. 8*, 2 (1969).
11. J. M. Prausnitz and R. Anderson, *AIChE J. 7*, 96 (1961).
12. R. F. Weimer and J. M. Prausnitz, *Hydrocarbon Process. Pet. Refiner 44*, 237 (1965).
13. P. A. Small, *J. Appl. Chem. 3*, 71 (1953).
14. C. M. Hansen and K. Skaarup, *J. Paint Technol. 39*, 511 (1967).
15. P. E. Froehling, D. M. Koenhen, A. Bantjes, and C. A. Smolders, *Polymer 17*, 835 (1976).
16. Z. Rigbi, *Polymer 19*, 1229 (1978).
17. P. E. Froehling and L. T. Hillegers, *Polymer 22*, 261 (1981).
18. P. J. Flory, *Principles of Polymer Chemistry*, Cornell University Press, Ithaca, N.Y. (1953).
19. M. L. Huggins, *Ann. N.Y. Acad. Sci. 43*, 1 (1942).
20. P. J. Flory, *J. Chem. Phys. 10*, 51 (1942).
21. W. R. Moore and R. P. Sheldon, *Polymer 2*, 315 (1961).
22. E. L. Lawton and D. M. Cates, *J. Appl. Polym. Sci. 13*, 899 (1969).
23. B. H. Knox, H.-D. Weigmann, and M. G. Scott, *Text. Res. J. 45*, 203 (1975).
24. A. S. Ribnick, H.-D. Weigmann, and M. G. Scott, *Text. Res. J. 42*, 720 (1972); *43*, 176 (1973).
25. A. S. Ribnick and H.-D. Weigmann, *Text. Res. J. 43*, 316 (1973).
26. H.-D. Weigmann and A. S. Ribnick, *Text. Res. J. 44*, 165 (1974).
27. H.-D. Weigmann, M. G. Scott, A. S. Ribnick, and L. Rebenfeld, *Text. Res. J. 46*, 574 (1976).
28. K. Suzuki, T. Asano, and I. Kido, *Sen-I Gakkaishi 30*, T143 (1974).
29. B. H. Knox, *J. Appl. Polym. Sci. 21*, 225, 249, 267 (1977).
30. W. Ingamells and N. Yanumet, *Brit. Polym. J. 12*, 12 (1980).
31. R. F. Blanks and J. M. Prausnitz, *Ind. Eng. Chem., Fundam. 3*, 1 (1964).
32. W. Ingamells and A. Yabani, *J. Soc. Dyers Col. 93*, 417 (1977).
33. W. Ingamells and A. Yabani, *J. Appl. Polym. Sci. 22*, 1583 (1978).
34. Z. Gur-Arieh, W. Ingamells, and R. H. Peters, *J. Appl. Polym. Sci. 20*, 41 (1976).
35. W. L. Lindner, *Colloid and Polym. Sci. 255*, 213 (1977).

36. W. R. Moore, *J. Soc. Dyers Col. 73*, 500 (1957).
37. A. Peterlin, *J. Macromol. Sci.—Phys. B11*, 57 (1975).
38. T. K. Kwei and T. T. Wang, *Macromolecules 5*, 128 (1972).
39. J. Crank and G. S. Park, Eds., *Diffusion in Polymers*, Academic, New York (1968).
40. S. P. Chu and J. D. Ferry, *Macromolecules 1*, 270 (1968).
41. C. A. Cumins and T. K. Kwei, *Diffusion in Polymers* (J. Crank and G. S. Park, Eds.), Academic, New York (1968).
42. H. Yasuda and A. Peterlin, *J. Appl. Polym. Sci. 18*, 531 (1974).
43. J. L. Williams and A. Peterlin, *J. Polym. Sci. A2, 9*, 1483 (1971).
44. A. Peterlin, *Pure Appl. Chem. 39*, 239 (1974).
45. Y. Takagi, *J. Appl. Polym. Sci. 9*, 3887 (1965).
46. G. S. Park, *Diffusion in Polymers* (J. Crank and G. S. Park, Eds.), Academic, New York (1968).
47. J. Crank and G. S. Park, *Trans. Faraday Soc. 47*, 1072 (1951).
48. T. Alfrey, E. F. Gurnee, and W. G. Lloyd, *J. Polym. Sci. C12*, 249 (1966).
49. H. B. Hopfenberg, V. T. Stannett, and G. M. Folk, *Polym. Eng. Sci. 15*, 261 (1975).
50. T. K. Kwei and H. M. Zupko, *J. Polym. Sci., A2, 7*, 867 (1969).
51. A. Peterlin, *J. Polym. Sci. B3*, 1083 (1965).
52. A. S. Michaels, H. J. Bixler, and H. B. Hopfenberg, *J. Appl. Polym. Sci. 12*, 991 (1968).
53. H. B. Hopfenberg, R. H. Holley, and V. T. Stannett, *Polym. Eng. Sci. 9*, 242 (1969).
54. A. Peterlin, *Makromol. Chem. 124*, 136 (1969).
55. T. Haga, *Sen-I Gakkaishi 34*, T282, T360 (1978).
56. P. J. Flory, *J. Chem. Phys. 17*, 223 (1949).
57. M. Gordon and J. S. Taylor, *J. Appl. Chem. 2*, 493 (1952).
58. R. G. Quynn, M. A. Sieminski, and J. L. Riley, *J. Text. Inst. 55*, T566 (1964).
59. L. Cottam, R. P. Sheldon, D. A. Hemsley, and R. P. Palmer, *J. Poly. Sci. B2*, 761 (1964).
60. A. Desai and G. L. Wilkes, *J. Polym. Sci., C46*, 291 (1974).
61. E. Balcerzyk, *Vysokomol. Soyed. A16*, 1581 (1974).
62. J. Kolb and E. F. Izard, *J. Appl. Phys. 20*, 571 (1949).
63. P. J. Makarewicz and G. L. Wilkes, *Text. Res. J. 48*, 136 (1978).
64. P. J. Makarewicz and G. L. Wilkes, *J. Polym. Sci. Phys. 16*, 1559 (1978).
65. G. H. Zachmann, *Faserforsch. Textiltech. 18*, 95 (1967).
66. G. H. Zachmann, *Makromol. Chem. 74*, 29 (1964).
67. G. M. Venkatesh, P. J. Bose, A. H. Kahn, J. P. Sibilia, and S. J. Hsu, *J. Appl. Polym. Sci. 26*, 223 (1981).
68. E. A. Gerold, L. Rebenfeld, M. G. Scott, and H.-D. Weigmann, *Text. Res. J. 49*, 652 (1979).

69. C. Durning, M. G. Scott and H.-D. Weigmann, paper in preparation.
70. K. Bredereck, H. Dolmetsch, E. Koch, and R. Schoner, *Melliand Textilber.* *56*, 50 (1975).
71. W. O. Statton, *The Setting of Fibers* (J. W. S. Hearle and L. W. C. Miles, Eds.), Merrow, Watford, England (1971).
72. P. F. Dismore and W. O. Statton, *J. Polym. Sci.* *C133*, (1966).
73. A. S. Ribnick, *Text. Res. J.* *39*, 428 (1969).
74. J. H. Dumbleton, *J. Polym. Sci.* *A2, 7*, 667 (1969).
75. A. S. Ribnick, *Text. Res. J.* *39*, 742 (1969).
76. M. P. W. Wilson, *Polymer* *15*, 277 (1974).
77. D. C. Prevorsek and A. V. Tobolsky, *Text. Res. J.* *33*, 795 (1963).
78. M. V. Forward and H. J. Palmer, *J. Text. Inst.* *41*, T 267 (1950); *45*, T 510 (1954).
79. M. I. Jacobs, L. Rebenfeld, and H. S. Taylor, *J. Appl. Polym. Sci.* *13*, 427 (1969).
80. M. I. Jacobs, *J. Appl. Polym. Sci.* *13*, 1191 (1969).
81. W. R. Moore, D. O. Richards, and R. P. Sheldon, *J. Text. Inst.* *51*, P438 (1960).
82. H.-D. Weigmann and A. S. Ribnick, paper in preparation.
83. B. H. Knox, Ph.D. Dissertation, Textile Research Institute and Dept. of Chemistry, Princeton University (1975).
84. D. C. Prevorsek, *J. Polym. Sci.* *C32*, 343 (1971).
85. D. C. Prevorsek, G. A. Tirpak, P. J. Harget, and A. C. Reimschuessel, *J. Macromol. Sci.—Phys.* *B9*, 733 (1974).
86. A. Peterlin, *Adv. Chem. Ser.* *142*, 1 (1975).
87. A. Peterlin, *J. Polym. Sci.* *C32*, 297 (1971).
88. J. L. Williams and A. Peterlin, *Makromol. Chem.* *135*, 41 (1970).
89. D. C. Hookway, *J. Text. Inst.* *49*, P292 (1958).
90. C. W. Bunn and T. C. Alcock, *Trans. Faraday Soc.* *41*, 323 (1945).
91. M. Ueda and Y. Nukushina, *Chem. High Polym.* (Japan) *21*, 337 (1964).
92. P. J. Makarewicz and G. L. Wilkes, *J. Polym. Sci., Polymer Phys. Ed.* *16*, 1559 (1978).
93. P. J. Makarewicz and G. L. Wilkes, *J. Appl. Polym. Sci.* *23*, 1619 (1979).
94. H.-D. Weigmann, M. G. Scott, and A. S. Ribnick, *Text. Res. J.* *47*, 761 (1977).
95. G. M. Venkatesh, A. H. Kahn, P. J. Bose, and G. L. Madan, *J. Appl. Polym. Sci.* *25*, 1601 (1980).
96. H.-D. Weigmann, M. G. Scott, and A. S. Ribnick, *Melliand Textilber.* *57*, 470 (1976).
97. H.-D. Weigmann, M. G. Scott, and A. S. Ribnick, *Text. Res. J.* *48*, 4 (1978).

98. K. H. Illers and H. Breuer, *Kolloid Z. 176*, 110 (1961).
99. M. Takayanagi, *Mem. Fac. Eng., Kyushu Univ. 23*, 65 (1965).
100. K. H. Illers and H. Breuer, *J. Colloid Sci. 18*, 1 (1963).
101. H. Jacobs and E. Jenckel, *Makromol. Chem. 47*, 72 (1961).
102. H. G. Zachmann and W. Schermann, *Kolloid Z., Z. Polym. 241*, 916 (1970).
103. E. A. Dimarzio and J. H. Gibbs, *J. Polym. Sci. A1*, 1417 (1963).
104. K. H. Illers, *Makromol. Chem. 38*, 168 (1960).
105. G. Goldbach, *Angew. Makromol. Chem. 32*, 37 (1973).
106. A. Desai and G. L. Wilkes, *Text. Res. J. 45*, 173 (1975).
107. J. P. Waters, *Text. Res. J. 41*, 857 (1971).
108. T. Murayama and A. A. Armstrong, *J. Polym. Sci., Polym. Phys. Ed. 12*, 1211 (1974).
109. E. L. Lawton and T. Murayama, *J. Appl. Polym. Sci. 20*, 3033 (1976).
110. D. L. McLean and T. Murayama, *Trans. Soc. Rheol. 18*, 237 (1974).
111. H. Fujita, A. Kishimoto, and K. Matsumoto, *Trans. Faraday Soc. 56*, 424 (1960).
112. J. H. Dumbleton, J. P. Bell, and T. Murayama, *J. Appl. Polym. Sci. 12*, 2491 (1968).
113. J. P. Bell, *J. Appl. Polym. Sci. 12*, 627 (1968).
114. H. H. Hofstetter, *Melliand Textilber. 50*, 321, 455, 845 (1969); *51*, 91, 336 (1970).
115. J. K. Skelly, *J. Soc. Dyers Col. 91*, 177 (1975).
116. K. Gebert, *J. Soc. Dyers Col. 87*, 509 (1971).
117. W. Kothe, *Proceedings of the AATCC Symposium Textile Solvent Technology: Update 1973*, Atlanta, p. 127 (1973).
118. F. O. Harris and T. H. Guion, *Proceedings of the AATCC Symposium Textile Solvent Technology: Update 1973*, Atlanta, p. 163 (1973).
119. F. O. Harris and T. H. Guion, *Text. Res. J. 42*, 626 (1972).
120. E. C. Ibe, *J. Appl. Polym. Sci. 14*, 837 (1970).
121. S. A. Siddiqui, *Text. Res. J. 51*, 527 (1981).
122. M. Forest, *Teintex 11*, 505 (1975).
123. B. Milicevic, *J. Soc. Dyers Col. 87*, 503 (1971).
124. J. Moreau and F. Somm, *Textilveredl. 7*, 211 (1972).
125. G. Reinert, W. Müller, R. Keller, and K. Schaffner, *Textilveredl. 7*, 678 (1972).
126. K. Bredereck and E. E. Koch, *Melliand Textilber. 55*, 157 (1974).
127. J. Rieker and N. Terlinden, *Melliand Textilber. 56*, 566 (1975).
128. B. L. Slaten, *J. Appl. Polym. Sci. 23*, 601 (1979).
129. T. K. Bhattacharje, *Amer. Dyestuff Reptr. 66* (9), 68, (1977).
130. V. C. Smith and F. E. Barwick, *Proceedings of the AATCC Symposium Textile Solvent Technology: Update 1973*, Atlanta, p. 135.

131. R. D. Matkowsky, H.-D. Weigmann, and M. G. Scott, *Text. Chem. Col. 12,* 55 (1980).
132. R. A. F. Moore and H.-D. Weigmann, *Text. Chem. Col. 13,* 70 (1981).
133. V. Zabashta, M. Sotton, and J. Jacquemart, *Bull. Scient. ITF 3,* 277 (1974).
134. T. V. Chambers and J. J. Hirshfeld (Milliken Res. Corp.), U.S. Patent 4,121,899 (1978).
135. W. Birke and H. U. von der Eltz (Hoechst), German Patents 2,720,586 and 2,720,661 (1978).
136. Y. Akiyama and K. Kimura (Kokai Tokkyo Koho), Japanese Patent 78,147,883 (1978).
137. H.-D. Weigmann, M. G. Scott, A. S. Ribnick, and R. D. Matkowsky, *Text. Res. J. 47,* 745 (1977).
138. S. P. McCarthy, L. Rebenfeld, and H.-D. Weigmann, *Text. Res. J. 51,* 323 (1981).
139. R. A. F. Moore and H.-D. Weigmann, paper in preparation.
140. R. A. F. Moore, H.-D. Weigmann, and C. A. Morello, *Text. Res. J.* (submitted for publication).
141. J. Mecheels, *Textilveredl. 4,* 749 (1969).
142. A. J. Shipman, *Textilveredl.* 5, 523 (1970).
143. G. Reinert, *Textilveredl. 10,* 85 (1975).

2

INTERACTION OF AQUEOUS SYSTEMS WITH FIBERS AND FABRICS

ISTVÁN RUSZNÁK Technical University of Budapest, Budapest, Hungary

1. Introduction 52

2. General Considerations 52

3. Sorption Processes in Textile Chemistry 55

3.1 Sorption influenced by the structure of fibers 55
3.2 General principles 58
3.3 Adsorption at the solid-liquid interface from various two-component systems 59
3.4 Adsorption at fiber-liquid interface 63
3.5 Interactions between fibers and adsorbed molecules 64
3.6 Sorption of disperse dyes by hydrophobic fibers 66
3.7 Sorption of resin monomers by cotton 67
3.8 Sorption of direct dyes by cellulose 68
3.9 Adsorption of polymers by fibers 70

4. Mass Transfer in Textile Chemical Processes 70

5. Penetration 72

5.1 Fiber penetration 72
5.2 Accessibility, reactivity, and stabilization of the swollen state of fibers 74
5.3 Prewetting followed by penetration of fabrics 76
5.4 Thermokinetics of liquid penetration into cellulose 85
5.5 Carrier supported penetration of fabrics 89

References 91

1. INTRODUCTION

The interactions of aqueous and fibrous systems are a major consideration in the chemical processing of fibers and fabrics. The complexity of the interactions is due partly to the complicated structure of fibers and fabrics, and also to the fact that most textile chemical processes are composed of a number of unit processes. Whether these processes occur sequentially or simultaneously depends upon the system itself and upon the process conditions.

In aqueous systems, at equilibrium prior to the interaction, unit processes follow each other according to a simplified model such as the following:

Wetting of the fiber surface
Diffusion of the solute in the aqueous solution toward the fiber surface
Start of the sorption of solute onto the fiber
Start of the penetration of solvent (water) and solute into the fiber
Start of fiber swelling
Diffusion of solute into the fibrous system
Equilibrium in the swelling process
Equilibrium in the diffusion process
Equilibrium in the sorption process
End of penetration
Distributional equilibrium of the solute between the bulk solution and the fibrous system

The occurrence and kinetics of all the unit processes mentioned above strictly depend upon the macroscopic, microscopic, submicroscopic, and molecular structure of the fiber. Due to this dependence, most of the processes are divided into four main groups [1]:

1. Surface processes involve only the segments of the fiber-forming polymer molecules located at the microscopic fiber surface.
2. Macroheterogeneous processes start at the accessible fiber surface and proceed layer by layer through the fibrous system as the fiber molecules dissolve or swell in the bulk solution.
3. Microheterogeneous processes swell the fiber-forming polymer, but the ordered or crystalline regions of the fiber do not become accessible to the solute.
4. Permutoid processes involve penetration of the solute and solvent molecules into both the open and the closed structural regions of the fiber, including fibrils and crystallites.

2. GENERAL CONSIDERATIONS

Almost all textile chemical processes consist of both a diffusion aspect and an affinity aspect. The first of these deals with the behavior of the dyestuff or chemical in solution and its movement into, within, or out of the fiber. The affinity aspect deals with the possible attachment of the solute molecules to the fiber-forming polymer molecules. It is not always easy to separate these two considerations in an ongoing process.

The diffusion aspect is of importance in determining the rate or the kinetics of the textile chemical process. The affinity aspect determines the total amount of solute taken up by the fibers upon attainment of true equilibrium between solute in the fibers and in solution.

Some of the solutes applied in textile chemical processes, especially certain dyes, exist in solution as aggregates of various sizes. During the exhaustion of the small aggregates onto the fibers, the larger aggregates break down to maintain an overall dynamic equilibrium of aggregates.

The molecules or aggregates of dyes or colloidal finishing chemicals are quite large, and it seems unlikely that they can penetrate the ordered crystalline microfibrillar regions of the various fibrous substances. Instead, they penetrate the fiber through pores and cavities present in the fibers, yarns, and fabrics. The swelling of the fiber makes penetration by larger molecules or aggregates easier. In spite of earlier assumptions, the particle size does not influence significantly the rate of textile chemical processes.

The degree of aggregation of solutes may be increased by the addition of electrolytes to a solution or colloidal dispersion. The opposite effect is attained by an increase in temperature; i.e., by increasing the kinetic energy of the molecules, existing aggregates are broken up into smaller units. Also, solute ions, molecules, or aggregates may associate with water molecules and become solvated. Thus, solutions of dyes and colloidal substances applied in textile chemical processes deviate considerably from the ideal characteristics of perfect solutions. This must be taken into consideration when attempting to interpret the thermodynamic properties of such solutions.

All the natural, and some of the synthetic, fibers immersed in water or aqueous solutions acquire positive or negative surface charges. Oppositely charged ions of dissolved dye or finishing-chemical molecules will be attracted toward the charged fiber, while ions carrying the same charge as the fiber surface will be repelled by the surface potential of the fiber. In the latter case, the potential barrier must be overcome before the solute ions can enter the fibrous system. This is highly dependent upon the energy content of the ions, which can be increased by raising the temperature of the solution; at higher temperatures, a greater proportion of the dissolved ions will be able to overcome the potential barrier and penetrate the fiber.

Neutral electrolytes added to the aqueous solution can also reduce the potential barrier. Ions can be attracted toward the oppositely charged fibrous surface in order to neutralize some of its repulsive effect on the ions of dye or finishing chemical.

Although, in most practical cases, true equilibrium cannot be reached in textile chemical processes, the assumption of equilibrium makes the interpretation of detailed processes easier. The solute-fiber system can be considered as reaching true equilibrium when each fiber has been evenly dyed or finished throughout its cross section and when no further change takes place in the concentration of the solute in the fiber as well as in the solution. In the initial state, the solute molecules are evenly distributed throughout a given volume of water in which untreated fiber, yarn, or fabric is placed. The solute molecules are in many cases ionized in the solution. The active ions of the solute, their counter ions, any ions of electrolyte present, and the water molecules are all in a state of motion. Some ions move faster than the others, and certain solute ions may possess enough energy to overcome the surface-potential barrier of the fibers when they are introduced into the system.

The strength of the attachment of solute ions to the fiber may not be large compared with the energy of the large dye or textile chemical ions; therefore these solute ions may become detached from a particular site and move to a new site inside the fiber. The stronger the forces of attachment of solute ions to fibers, the more difficult they will be to detach. This accounts for the well-known observation that dyes and textile finishing agents differ in their wet-fastness properties or in their substantivity toward the fiber.

In solute-fiber systems at equilibrium, solute molecules, free to move independently in the solution, become concentrated into a much smaller volume in the fiber. This requires some motivating force, i.e., the affinity of the solute for fiber. The chemical potentials of the two phases, solute in solution and solute in fiber, can be calculated. At the start of textile chemical processes the chemical potential of the solution phase exceeds that of the fiber phase. When equilibrium is reached, the chemical potentials of the two phases will be equal.

It is difficult to determine the available volume within the fibers into which the solute molecules may pack. The manner in which the solute particles arrange themselves in this space affects the calculation of the chemical potential of the solute in the fiber phase. Three schemes of arrangement may be considered:

1. Nonionized solute yielding a "solid solution" in the fiber
2. Solute adsorbed on specific sites at the fiber-solution interfaces in a monomolecular layer
3. Solute attracted to the fiber-solution interfaces without being adsorbed on specific sites, i.e., capable of forming multimolecular layers (diffuse adsorption)

These schemes may be used to derive adsorption isotherms for any particular solute-fiber system.

The two types of adsorption, multilayer and monolayer, are the two extremes of a common phenomenon. The type of adsorption which occurs depends upon the magnitude and sign of the charges on the fiber surface and solute ions. Diffuse adsorption is brought about by zero surface potential on the fiber; monomolecular solute ion distribution is the consequence of a high fiber-surface potential. The fiber surface should be considered as including all the internal surfaces of the pores and cavities in the fiber which are accessible to the aqueous solution as the fiber swells during textile chemical processes.

3. SORPTION PROCESSES IN TEXTILE CHEMISTRY

When the concentration of an active component in a fibrous system changes relative to the bulk solution, sorption processes are operating. For instance, the dyeing of a fiber is usually caused by adsorption; the loss in dyestuff during rinsing is caused by desorption.

The internal surface of a fiber is more important than the external surface in determining whether or not a fiber is a good adsorbent. To become adsorbed, molecules must come into contact with active surfaces within the pores and with the free volume of the fibers. Free volume is generated by the movement of fibrous polymer molecules and varies continually with respect to location and size. Consequently, the possible active internal surfaces are the pore walls, the polymer chains next to the free volumes, the surface of the crystallites, and any other accessible molecular segments or sites. For molecules to be adsorbed on a fiber surface, the two entities must come within a distance of less than 0.5 nm in order to enable primary or secondary forces to generate adsorption.

3.1 Sorption Influenced by the Structure of Fibers

Fibers are solid adsorbents. Surface polymer molecules are influenced by anisotropic forces, since they are surrounded by similar molecules only from the internal part of the solid phase. Thus, attractive forces generated on them are directed toward the center of the solid phase (Fig. 2.1).

The free-energy content of a molecule on the surface is higher than that of a molecule in the bulk phase. Thus, work must be done to carry a molecule from the bulk phase to the surface of the fiber. The specific free surface energy (SFSE) is the work necessary for carrying as many molecules to the surface as is necessary for the formation of 1 cm^2 of new surface (Table 2.1).

The thermodynamic tendency to decrease the free-energy content of a system to a possible minimum value can be achieved through the sorption of molecules, aside from the other polymer molecules, on the surface of the fiber. The higher the SFSE of an adsorbent, the larger is the surface of its mass unit, which is measured in units of m^2/g. The SFSE values of dry fibers are not too large. Data calculated from nitrogen sorption by BET calculations are shown in Table 2.2.

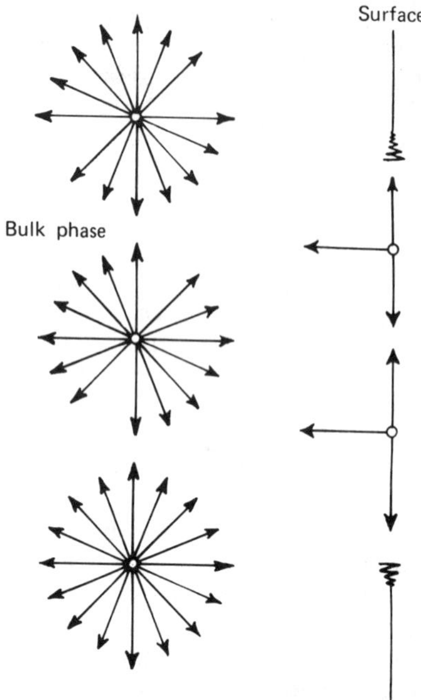

Figure 2.1 Forces acting on molecules situated in the bulk phase and on the surface of the fiber. Adapted from Ref. 2.

Fibers immersed in certain liquids show much increased surfaces. For instance, for cotton a 400 m^2/g surface has been measured. Figures are valid only for selected pairs of fibers and liquids.

Heat is evolved during the wetting of a fiber by water. The so-called heat of immersion can be used as a measure of active surface area (Table 2.3).

Table 2.1 Specific Free Surface Energy of Fibers

Fiber	SFSE (μJm^{-6})
Viscose rayon	2.9
Polyamide	3.0
Acrylics	3.7
Polyester	4.2

Source: Adapted from Ref. 3.

Table 2.2 Active Surface Area of Fibers

Substrate	Area (m^2/g)
Cotton	0.72
Wool	0.96
Polyamide	0.31
Cellulose triacetate	0.38
Acrylics	1.00
Polyester	0.50
Activated charcoal	200-1500

Source: Adapted from Ref. 4.

Hydrophilic and hydrophobic fibers are characterized by polar and nonpolar groups, respectively, in their polymer molecules (Table 2.4). The hydrophilic nature of a fiber can be estimated by its cross-sectional swelling in water. A lack of swelling can be caused either by the absence of polar groups (e.g., as in polyester) or by their nonaccessible location (e.g., as in cellulose triacetate).

The role of crystallites in adsorption is determined by their size, type, orientation, and distribution. Other influencing factors are the presence and size of amorphous regions or voids within them; their shape, density, and total volume; and intermolecular

Table 2.3 Heat of Immersion of Fibers in Water

Fiber	Heat of immersion (J/g)
Cotton (bleached)	46.2
Cotton (mercerized)	73.5
Viscose rayon	92.4
Linen	54.6
Wool	113.0
Polyamide	30.7
Polyester	5.0
Acrylics	7.1

Source: Adapted from Ref. 5.

Table 2.4 Hydrophilicity of Fibers

Fiber	Groups	Swelling in water (% increase in diameter)
Cotton	-OH	44-49
Viscose rayon	-OH, -COOH	45-82
Wool	-CO-NH-, -COOH, $-NH_2$	32-38
Secondary cellulose acetate	-OH, $-O-\overset{\displaystyle O}{\overset{\|}{C}}-CH_3$	6-30
Cellulose triacetate	$-O-\overset{\displaystyle O}{\overset{\|}{C}}-CH_3$	-
Polyamide	$-NH_2$, -COOH, -CO-NH-	2
Acrylics	-CN, -COOH, $-SO_3H$, $-OSO_3H$	-
Polyester	-OH, -COOH, -O-CO-	-

Source: Adapted from Ref. 6.

orientation and interactions. The glass transition temperature, which is dependent on crystalline and amorphous structure, is another indicator of the adsorption process on fibers. The vibrational movement of molecular segments increases at this temperature, enabling the migration of adsorbed molecules into the polymer structure.

3.2 General Principles

Adsorption from solution can occur at one of the interfaces of the liquid with a vapor, a solid, or another liquid. Many papers have been published on adsorption from solution under equilibrium conditions and with adsorption isotherms. However, only a few studies have been published on adsorption at different temperatures. The kinetics of adsorption, as well as the dynamic aspect of the process, are equally important when fibrous adsorbents are used.

Adsorption, especially from solution, is mainly due to the effect of physical forces. On a number of solid adsorbents, however, it can be shown that chemisorption occurs from proper solutions. Chemisorption from solution may be accompanied by physical adsorption onto the chemisorbed layer. Similarities as well as differences can be found between the adsorption of electrolytes and nonelectrolytes. Large, technically important ions — such as dyestuffs, surfactants, and polyelectrolytes — represent an intermediate position. Their adsorption is predominantly due to the effect of van der Waals forces, because the ion is large relative to the magnitude of its charge. Adsorption processes of this type of electrolyte and nonelectrolyte show many similarities.

Due to differences in adsorption shown by various two-component systems, their classification is important. Among them are the following:

1. Two completely miscible liquids
2. Liquid in liquid, with partial miscibility
3. Solid in liquid, with partial miscibility
4. Gas in liquid

Adsorption on a solid-liquid interface is the most suitable model to represent sorption processes on fibrous adsorbents.

3.3 Adsorption at the Solid-Liquid Interface from Various Two-Component Systems

Adsorption, according to a simple definition, describes the concentration of one component of a system at an interface relative to its concentration in the adjacent bulk phase. Experimental data are mainly interpreted by means of adsorption isotherms, plotting concentration (or pressure) against adsorption. Isotherms can be fitted by different equations. The least complicated, the Freundlich equation, can be applied, with restrictions, to adsorption of gases. The Langmuir and other not too complicated equations have been proposed, but none of them (except the Gibbs equation, which involves interfacial tension) can yet be applied universally for all kinds of adsorption isotherms. The quantity of the gas adsorbed by the unit weight of a solid is demonstrated by the adsorption isotherms of a single gas by a solid adsorbent. A change in concentration occurs, however, when a solute is adsorbed by a solid adsorbent from its solution, assuming that the solvent is not adsorbed at the same time. This is true for solutes having very limited solubility in the solvent.

In systems which include more soluble (or completely miscible) solutes, each component has to be regarded as a possible adsorbate. Many examples can be found among fibrous adsorbents to demonstrate that water might be considered in an aqueous solution not only as a solvent but also as a possible adsorbate. Positive adsorption of the first component by an adsorbent from a two-component system means that it is adsorbed preferentially with respect to the second one. Preferential adsorption of the second component, however, means a negative adsorption of the first one. This is the case of preferential adsorption at the liquid-solid interface, corresponding with the term of surface excess, applied for the same type of adsorption at the liquid-vapor interface. Absolute adsorption refers to the quantity of an individual component in the adsorbed phase. Preferential adsorption from a two-component system is represented by a composite isotherm, i.e., by the combination of the individual isotherms for the adsorption of each component.

No suitable equation has been developed for the description of the composite isotherm of real systems. Such an isotherm can, however, be considered as one consisting of two individual isotherm components. A single molecular layer is the adsorbed phase in the first component, and a pore-filling phase is the second component. The monolayer adsorbed phase consists of both components, but, due to preferential adsorption, its composition is not the same as that of the bulk two-component liquid. This first phase is governed by the available surface of the adsorbent, while the available pore volume controls the pore-filling process

The adsorption of only one component, which may be due to the molecular sieve action of the porous solid adsorbent, can be considered as a special case of pore-filling adsorption. Monolayer adsorption may be followed by capillary adsorption, i.e., mechanical filling of the pores by the bulk liquid in the case of solid adsorbents with rather wide pores.

The four main types of solid adsorbents are as follows:

Nonporous adsorbents
Adsorbents with pores wider than the average molecular diameters
Adsorbents with pores narrower than the average molecular diameters
Adsorbents acting as molecular sieves

Solid adsorbents may overlap each other slightly in their characteristics. Fabrics can be considered as a combination of the last three types of solid adsorbents.

Monolayers adsorbed from solution usually contain a mixture of all the components of the solution. Adsorption very often does not stop there, but builds up further adsorbate layers. The more layers that are included in the multilayer adsorbate phase, the nearer is the last layer in its composition to that of the bulk solution. In special cases the adsorbate phase consists of two layers, the first of which is chemisorbed, while the second is physically adsorbed. In certain cases, especially when water is one component of the bulk liquid, such a monolayer adsorption may occur in which complex (e.g., dimeric) molecules form the adsorbate phase on the surface of the adsorbent.

The adsorption of water vapor in fibrous adsorbents is represented by isotherms of sigmoid shape, which can be considered as the superposition of two separate curves (Fig. 2.2). Curve I represents the directly sorbed amount of water vapor, which does not change after the formation of a strongly bound monomolecular layer. Curve II

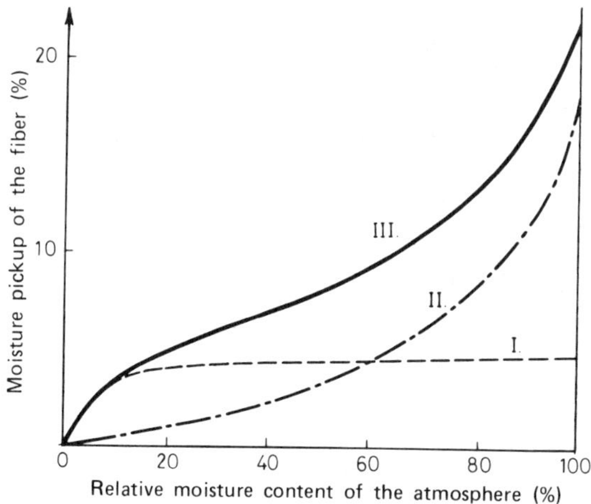

Figure 2.2 Composite water vapor sorption isotherm of a fibrous adsorbent.

represents the loosely bound condensation of further water onto the monomolecular water layer. The isotherm, curve III, is the sum of the two components [7]. Only the second component can be considered as being fully reversible in the course of desorption.

The strength of the bond between fiber and water molecules can be characterized by the released heat of sorption. This is equal

Table 2.5 Differential Heat of Sorption of Native Cellulose as a Function of its Moisture Content

Moisture content (%)	Differential heat of sorption (J/g water)
0.5	1344
1.0	1152
2.0	912
4.0	720
6.0	528
8.0	384
10.0	240
12.0	144

to the difference between the enthalpies of the free and of the sorbed vapor. Direct determination is not yet established. However, it is simple to measure the heat of wetting of a dry sample Q_1 and of a sample Q_2 containing 5% adsorbed moisture (Table 2.5).

The relationships concerning the sorption of water on polymers and fibers are relatively complex and can be resolved by a number of definitions which follow:

1. Q_L = the differential heat of sorption of liquid water, i.e., the amount of heat released when a unit quantity of liquid water is adsorbed by an infinite quantity of polymer of regain α. It is expressed in Cal/g of liquid water.
2. Q_V = the differential heat of sorption of water vapor, as above. It is expressed in Cal/g of water vapor. It can be seen that $Q_V = Q_L + L$, where L is the latent heat of condensation of water vapor, i.e., 582 Cal/g at 25°C.
3. H_L = the integral heat of sorption of liquid water, i.e., $H_L^{0-\alpha}$, the amount of heat released from a dry polymer upon absorption of an amount of water sufficient for raising the regain of 1 g of polymer from 0 to α. It is expressed in Cal/g of dry polymer.
4. H_V = the integral heat of sorption of water vapor, i.e., $H_V^{0-\alpha}$. It is related to H_L by the equation

$$H_V = H_L + \alpha L$$

5. $\overline{W}$ or $H_L^{0-\alpha_s}$ = the total heat of wetting, i.e., the heat released when 1 g of dry polymer absorbs the quantity of water needed for saturation. α_s = the regain at saturation.
6. W or $H_L^{\alpha-\alpha_s}$ = the partial heat of wetting, i.e., the heat released when 1 g of dry polymer of regain $\underline{\alpha}$ absorbs the quantity of water needed for saturation. $\overline{W}$ and W are expressed in Cal/g of dry polymer.

It can easily be seen that

$$H_L^{0-\alpha} + H_L^{\alpha-\alpha_s} = H_L^{0-\alpha_s} = \overline{W}$$

$$H_L = \overline{W} - W$$

$$\frac{dH_L}{d\alpha} = \frac{-dW}{d\alpha} = Q_L$$

and

$$W = \int_{\alpha}^{\alpha_s} Q_L d$$

Q_L can be obtained from the plot of W against α by measuring the slope of the tangent at a given value of α [8]. The heat of wetting can be determined by calorimetry. Q_L decreases with the increase in α. It measures the energy required to release absorbed water molecules from the polymer, i.e., the strength of the bond between the water molecule and the polar site to which it is attached. Since it is determined by the nature of the polar site, it characterizes a given polymer. The integral heat of wetting $\overline{W}$, however, depends largely on the quantity of available polar sites in the polymer mass, i.e., on the amount of noncrystalline matter.

If the porous adsorbent acts as a molecular sieve, only one component of the bulk liquid can be adsorbed. Preferential physical adsorption decreases with increasing temperature, while chemisorption does not usually change.

Isotherms can be used for the comparison of different adsorbents, providing that the amount of preferentially adsorbed substrate is plotted against the active surface area of the adsorbent.

In multilayer adsorption, as already mentioned, the first step may often be a slow chemisorption, altering the surface of the adsorbent, followed by a physical adsorption on the new surface. In other cases, the above process can be preceded by a physical adsorption on the original surface. Chemisorption is usually combined with the preferential physical adsorption of one component of the bulk liquid.

3.4 Adsorption at Fiber-Liquid Interface

The adsorption onto fibers of solid solutes from their solutions is an important consideration. The most frequently occurring isotherm (Figure 2.3) approaches a limiting value. The resulting plateau can be attributed to the formation of a complete solute monolayer on the adsorbent.

For a complete representation of the adsorption process, however, the possibility of solvent adsorption by the fiber should not be completely neglected. This might be evidenced on some isotherms by a further rise beyond the plateau or a definite maximum and a subsequent declining section. Sometimes a mixed monolayer containing solute and solvent in constant composition can be formed on the adsorbent [9]. The absence of a limiting value in the isotherm may be due either to crystallization of the solute or to multilayer formation.

Due to the increased solubility of the solute in the solvent and to the diminishing attraction between the solute molecules and the adsorbent, a rise in temperature depresses the limiting adsorption value from solutions of low concentration but does not affect the limiting adsorption value from solutions of high concentration. As the temperature of the system rises, part of the adsorbed solute is redissolved in the bulk solution.

If the sorption from the solution of solid solute is a chemisorption phenomenon, the adsorbed amount of the solute either will be independent of the concentration of the bulk liquid (i.e., no accompanying

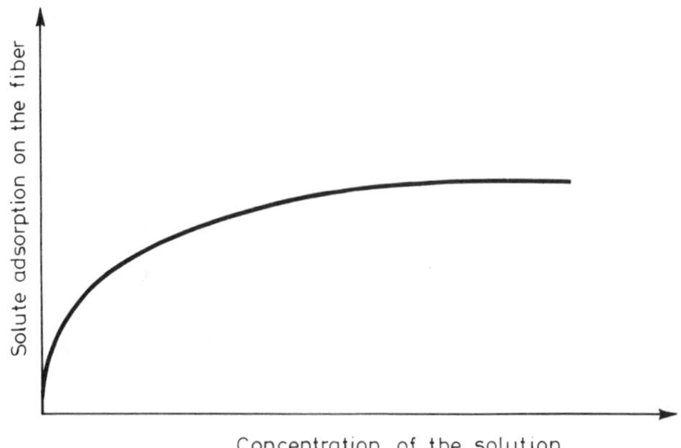

Figure 2.3 Adsorption isotherm of a solid solute on a fibrous adsorbent.

preferential physical adsorption) or will increase with the increase in concentration of the bulk liquid (i.e., chemisorption with simultaneous preferential physical adsorption). In both cases the isotherm starts above the origin of the diagram.

3.5 Interactions Between Fibers and Adsorbed Molecules

The difference in the chemical potentials of a solute, e.g., a dye, in a fiber in the standard state (i.e., 1 mol of dye per kg of fiber) and in a solution in the standard state (1 mol of dye per 100 cm^3 of solution) has been defined as the normal affinity or substantivity:

$$\mu_0 = \mu_0^0 + RT \ln(c)_0$$

$$\mu_\phi = \mu_\phi^0 + RT \ln(c)_\phi$$

where

μ_0 = chemical potential of the solute in solution

μ_0^0 = the same as μ_0 in standard state

μ_ϕ = chemical potential of the solute in the fiber

μ_ϕ^0 = the same as μ_ϕ in standard state

μ_0 = μ_ϕ in equilibrium

$$-\Delta\mu^0 = -(\mu_0^0 - \mu_\phi^0) = RT \ln \frac{(c)_\phi}{(c)_0}$$

Normal affinity equals the heat of sorption (ΔH^0) minus the entropy of sorption (ΔS^0):

$$\Delta\mu^0 = \Delta H^0 - T\Delta S^0$$

The difference between the amount of heat necessary for the removal of 1 mol of dye from the solution and for the removal of 1 mol of dye from the fiber can be defined as ΔH^0. This is equal to the previously defined differential heat of sorption.

The entropy of sorption is the ratio of the probability of sorption (β_1) to that of desorption (β_2) and μ_0 is the resultant of forces establishing order and disorder. ΔH^0 can be calculated from affinity values determined at several temperatures:

$$\Delta H^0 = \frac{d(\Delta\mu^0/T)}{d(1/T)}$$

or

$$\frac{\Delta H^0}{T} = \frac{\Delta\mu^0}{T} + \Delta S^0$$

In the plot of $\Delta\mu^0/T$ against $1/T$, ΔH^0 can be calculated from the slope of the line. Thus, knowing ΔH^0, ΔS^0 can be calculated (Fig. 2.4). In most cases, the increase in temperature diminishes the affinity of the sorbed dyestuff.

The adsorbed amount of solute per mass unit of adsorbent decreases exponentially with the amount of adsorbent;

$$(c)_\phi = \frac{(c)_{0,0}}{m}[1 - \exp(km)]$$

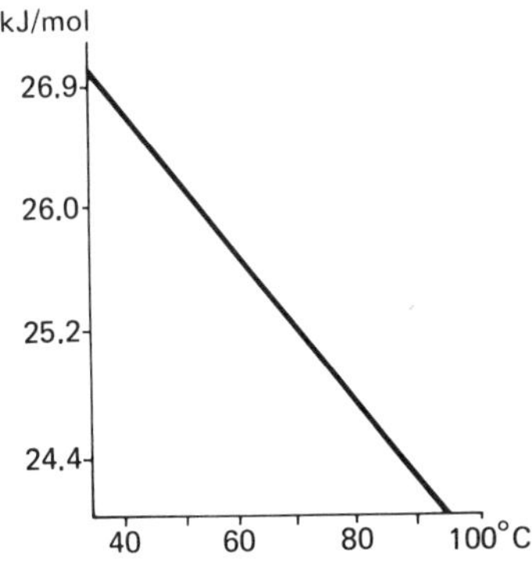

Figure 2.4 Change in the affinity of a disperse dye with temperature. (Adapted from Ref. 10.)

where

$$k = \text{constant}$$

$$c_{0,0} = \text{initial concentration of the solution (g/cm}^3\text{)}$$

$$m = \text{the mass of adsorbent (g)}$$

$$(c)_\phi = \text{g adsorbate per g adsorbent}$$

3.6 Sorption of Disperse Dyes by Hydrophobic Fibers

It can be demonstrated that single molecules of disperse dyes penetrate the fiber structure. Interaction is based on van der Waals forces and on hydrogen bonds. The amount of disperse dye adsorbed by a hydrophobic fiber increases until a certain saturation value is reached. The sorption isotherms of disperse dyes on hydrophobic fibers are of the Nernst type, such as that shown in Fig. 2.5. The correlation

$$\frac{(c)_\phi}{(c)_0} = K$$

where K is the distribution constant, describes the adsorption process up to the inflection point of the isotherm.

The saturation value is not due, however, to a limit in the dissolving capacity of the fiber, since it can be demonstrated that the limit in adsorption of disperse dyes on hydrophobic fibers is due to the limited solubility of the dyes in water and not to their dissolution in the fiber. The increase in concentration of the dispersion does not change this limiting value. The increase in solubility, however, increases the concentration in the fiber (Fig. 2.6)

A close correlation has been determined between wet-fastness characteristics and affinity of the dye to the fiber:

$$-\Delta\mu^0 = RT \ln \frac{(c)_\phi}{(c)_0} = RT \ln K$$

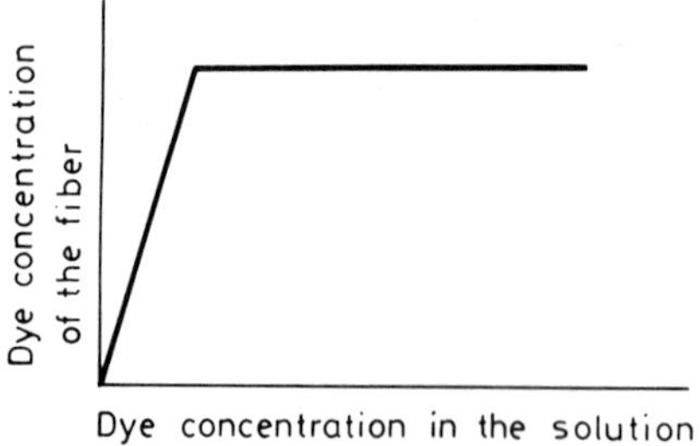

Figure 2.5 Sorption of a disperse dye by a hydrophobic fiber. (Adapted from Ref. 11.)

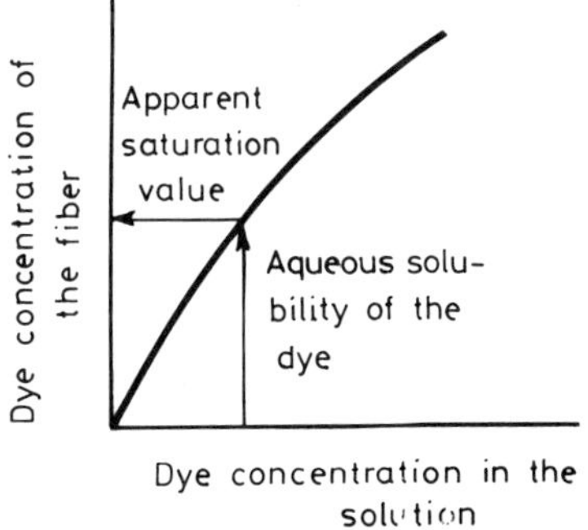

Figure 2.6 Sorption of a disperse dye by a hydrophobic fiber, influenced by the limited solubility of the dye. (Adapted from Ref. 11.)

The greater the distribution constant, the higher the affinity of the dye. The distribution constant decreases with an increase in temperature (Table 2.6). In spite of this, the maximum in the adsorbed amount of dye increases due to the increased water solubility of the dye.

The sorption process of a given disperse dye is not influenced by the presence of another dye of the same chemical structure (e.g., azoic or anthraquinoid) in the system. A deeper shade, however, can be achieved by applying a blend of two different disperse dyes of similar chemical nature in the same aqueous dyebath dispersion than with either of the two separately.

Heat treatment of synthetic fibers brings about a diminished sorption of disperse dyes.

3.7 Sorption of Resin Monomers by Cotton

A sorption process precedes cross-linking reactions in cotton. Measurements carried out with the DMEU-cotton system at different temperatures (20, 30, 40, and 50°C) have shown that an increase in $(c)_{\phi}$

Table 2.6 Effect of Temperature on Affinity

Dye	Affinity (kJ/mol) 100°C	Affinity (kJ/mol) 150°C	H^0 (kJ/mol)	S^0 (J/mol °C)
C.I. Disperse Blue 152	24.90	21.42	-51.16	-70.35
C.I. Disperse Yellow 54	28.77	25.83	-50.74	-58.93

occurred at each temperature at the beginning of the sorption process. Subsequently, its value suddenly drops, and finally the value of $(c)_\phi$ gives evidence of negative adsorption. The process can be represented by the equation

$$\frac{1}{(c)_\phi} = \frac{1}{K(c_\infty)_\phi (c)_0} + \frac{1}{(c_\infty)_\phi}$$

where

K = equilibrium constant

$(c)_\phi$ = mols of DMEU adsorbed by 1 kg cotton

$(c)_0$ = mols of DMEU in 1 cm^3 solution

$(c_\infty)_\phi$ = saturation limiting value of $(c)_\phi$

Through a linear correlation between $1/(c)_\phi$ and the corresponding $1/(c)_0$ values, the validity of a Langmuir-type isotherm on the equilibrium adsorption of DMEU by cotton has been demonstrated. Affinity values calculated from the slopes of the isotherms were used for the calculation of enthalpy and entropy parameters of the adsorption process (Table 2.7).

The accessible surface area of cotton to DMEU has been calculated from the area demand of a single molecule as 111 m^2/g. The negative adsorption, occurring after an increase in the concentration of DMEU, is the consequence of the preferential adsorption of water by cotton.

From the shape of the adsorption isotherm, the occurrence of associated DMEU molecules can be inferred. It could be demonstrated that aqueous DMEU solutions contain approximately equal quantities of both single and aggregate molecules, as well as their protonated derivatives.

Monomolecular adsorption is the first step of the sorption process; multilayer adsorption of the sorption of aggregates are equally excluded. Concentrations exceeding a critical level lead to the formation of aggregates of DMEU in the treating solution, and thus negative adsorption occurs. Dimers show no affinity toward cotton. The adsorbed amount of water, under these conditions, is nearly equal to that which is required for complete monolayer adsorption on cotton. Due to the strong bond of monomer molecules to the fiber, dimer DMEU molecules are not able to replace them.

3.8 Sorption of Direct Dyes by Cellulose

Allingham [14] gave evidence against the role of H bonds in the direct dyeing of cellulose. It could be demonstrated that molecules wandering in the water-filled pores of cellulose are bound to the walls of crystallites by means of van der Waals interactions. The dye molecules

Table 2.7 Temperature Dependence of Affinity, Enthalpy, and Entropy of DMEU Adsorption on Cotton

Temperature (°C)	Affinity (kJ/mol)	Enthalpy (kJ/mol)	Entropy (J/mol°C)	Area demand (Å^2/mol)
20	-10.00	-21.13	-37.38	40
30	-9.66	-19.87	-33.60	40
40	-9.32	-19.82	-33.60	42
50	-8.95	-19.78	-33.60	43

Source: Adapted from Ref. 13.

are adsorbed separately, but form a surface excess and then associate within the fiber.

A Langmuir-type sorption isotherm describes the above process until a given dye concentration (corresponding to a monomolecular layer) is reached. After this, a sudden increase in dye concentration in the fiber occurs. The dye concentration corresponding with the inflection of the isotherm is the critical micelle concentration (cmc) value of the dye.

The negative charge on the cellulose surface, caused partly by the presence of -COOH groups, increases the repulsion forces on the anionic dye molecules. Due to this, an apparent saturation value smaller than the real one is sometimes determined.

The affinity of cellulose for direct dyes decreases with increasing temperature. The addition of electrolyte may increase the amount of dye adsorbed by cellulose due to a reduction in the electrokinetic potential (negative charge) of the cellulose surface.

Two different direct dyes, when adsorbed from the same solution, compete for the accessible sites of cellulose, thus diminishing the adsorption values of each other. The distribution between them on the cellulose occurs according to the ratio of their affinity values.

3.9 Adsorption of Polymers by Fibers

Polydispersity is a general characteristic of polymers. Molecules of different degrees of polymerization are adsorbed at different rates. Adsorption usually increases with increasing $\overline{DP}$ of the polymer adsorbate, which is associated with its decreasing solubility in the solvent. Especially near the critical miscibility temperature, the solubility of different polymer fractions is highly dependent upon their molecular weight. An increase in the solubility parameter of the solvent brings about a decrease in adsorption, unless the effect is compensated for by a strong interaction between polymer and adsorbent.

An increase in temperature often has a negative effect on adsorption of polymers. This may be negated, however, if the increase in temperature improves the solubility of the polymer, thus causing an increase in concentration in the bulk liquid.

Adsorption of dissolved polymers also appears to be limited by the viscosity of the solution of interest. The higher the molecular weight or viscosity of a polymer fraction, the less likely it is that it can penetrate easily into a porous adsorbent.

4. MASS TRANSFER IN TEXTILE CHEMICAL PROCESSES

In kinetic studies of textile chemical processes, the assumption has generally been made that intensive agitation of the liquid system is sufficient to enable the solute molecules to reach the fiber surface in an extremely short time. The time required is significantly longer in stagnant liquid systems. It has been shown that the time of half-dyeing in a cotton-direct dye system at 60°C could be reduced from 10-15 min to 2-5 min by raising the intensity of the agitation applied 10-fold [14].

Due to nonuniform accessibility of fiber surfaces, stagnant regions can occur in which there is no relative motion between fiber and solution. The probability of the occurrence of such regions is greatest in fabrics and knit goods, less in yarns, and least in polymer films. They form mainly around points of contact between fibers and yarns. A 10-fold increase in the intensity of agitation brought about a 10% increase in the dye uptake of films or yarns and a 50% increase in that of fabrics.

In the course of relative motion between fibers and solution, an unmoving liquid layer adheres to the fiber surface. The distance between the fiber surface and the moving bulk liquor can be considered as the thickness of the adjacent boundary layer. The thickness of the boundary layer is inversely proportional to the square root of the rate of the relative motion, i.e., the rate of liquor flow. Under similar conditions, the initial thickness is higher on fabrics than on films; its value is highly dependent upon the fineness of the yarns and upon the structure and density of the fabrics.

In the course of acid dyeing of wool, a correlation could be found [14] between the amount of dissolved dyestuff (J_s) diffusing across the boundary layer, the diffusion coefficient of the dye in solution (D_s^0), the active surface area of the fiber (A), the dye concentration in the solution (c_s), and the thickness of the boundary layer (d), as follows:

$$J_s = D_s^0 A \frac{c_s}{d}$$

The amount of dye moving across the boundary layer in a given time unit is equal to the amount of dye penetrating the fiber within the same time if the rate of penetration is higher than that of dye diffusion in the solution. Based upon the above equation, the thickness of the boundary layer can be calculated.

Textile chemical solutes are subjects of two different transfer processes. The first is determined by the velocity at which its solution is swept past and the second is determined by the rate at which the solute diffuses from its solution to the surface of the fiber. It can be demonstrated that the diffusional boundary layer is much thinner than the hydrodynamic boundary layer.

Sorption and penetration processes are diffusion controlled if the bath is infinite. This happens when the concentration of solute is low in the fiber and the boundary layer is thick (i.e., the rate of flow is low). An increase in the concentration of solute within the fiber makes the process *fiber controlled* (i.e., dependent upon the diffusion of the solute into the fiber).

5. PENETRATION

Dissolved colored or colorless compounds should approach active sites of the fibers in most textile chemical processes and operations. The initial part of this rather complex mass transport is penetration. Penetration can be considered as the teamwork of the wetting, sorption, diffusion, and swelling processes. It is not surprising, consequently, that penetration is greatly influenced by all the parameters and characteristics of the fibers, yarns, and fabrics, as well as by those of the other constituents of the system. Fibers and filaments can be considered in that respect as relatively simple systems, while woven and knitted fabrics are the most complex systems.

5.1 Fiber Penetration

After arrival at the active sites of the fibers, dissolved ions, molecules, or particles of chemical compounds form bonds with reactive sites in the fiber or enter into in situ reactions with other compounds. Compounds within the fibers are much more protected against the effects of laundering, abrasion, and light than those on the fiber surface. The fiber surface cannot offer enough sites for binding reactive dyes or finishing agents, and there can be no efficient reactive dyeing or finishing without fast and uniform penetration. The ions, molecules, or particles of the penetrating compounds should find enough space between the adjacent polymer molecules to reach the accessible sites. The submicroscopic structure of the fibers and changes in it brought about by swelling, temperature of the system, applied pressure, time available for swelling, or wet processing are a few of the parameters which influence the accessibility.

With hydrophobic fibers, molecules or particles of flame retardant, water repellent, soil release, stiffening, or antistatic finishing agents penetrate mostly between the fibers and are fixed mainly on their surfaces. Insolubilization of certain finishes only around the fibers, instead of in them, can lead to some macroscopic effects, such as harsh hand or reduced durability.

All compounds which are preferentially adsorbed by the substrate in the course of penetration of the fibers can be considered substantive. Three mechanisms of fiber penetration are known [1]:

1. Penetration into the existing pores of the fibers
2. Penetration preceded by swelling of the fibers
3. Penetration during simultaneous swelling of the fibers

The voids or pores encountered in fibers are isolated from one another. A system of interconnected voids does occur, however, in yarns and fabrics. Thus the penetration of fibers brings about mostly swelling of the system. Additivity of fiber and liquid volumes seldom occurs. A negative difference from additivity is brought about by contraction and a positive one by expansion of the penetrated fibers.

Molecules of the solvent penetrate the delocalized voids of the fibers with the occurrence of simultaneously generated new voids according to the so-called free-volume model of penetration. A good way to follow fiber penetration is to assume that the water molecules precede the solute molecules in penetrating the hydrophilic fibers and generate, through swelling, an interconnected pore system out of the previously isolated voids. This pore system, however, is not stable to drying unless a solvent exchange has been applied.

Fibers, considered as submicroscopically double-phase systems, are penetrated by solvent and solute molecules first in areas of least lateral order. Consequently, an increase in distance between adjacent segments occurs in these regions. These swollen regions within the fibers are interspaced with unchanged crystallites. The smaller the distance between two crystallites, the less effective is the swelling of the fiber. Aqueous swelling of hydrophilic fibers is a highly exothermic process, while very little change in heat content of the system occurs in the course of aqueous treatments of hydrophobic fibers.

Attractive forces between cellulose and water are much stronger than the ones among water molecules only. Accordingly, water is adsorbed by cellulose. The value of the differential heat of adsorption (the amount of heat liberated in the course of adsorption of 1 g of water by cellulose of infinite mass) can be taken as the measure of attraction between cellulose and water. Its value is constant for all kinds of cellulosic fibers, while the values of heats of wetting vary depending upon the submicroscopic structure and accessibility of the substrate [15]. For instance, cotton exhibits a heat of wetting of 42 J/g; mercerized cotton, 62 J/g; and viscose rayon, 105 J/g.

The fibrillar structure of cotton and some other fibers means a nearly periodic fluctuation in density and thus in the accessibility of their variously ordered regions. Penetration into the fibrillar structure is limited mostly to the surface of microfibrils. Regions of maximum lateral order remain unchanged in the course of interfibrillar swelling in most wet textile chemical processing operations. Intrafibrillar swelling, with the exception of a few processes such as mercerization and liquid ammonia treatment of cotton, is unlimited and it is mostly followed by dissolution of the fiber.

The closer the solubility parameter (square root of the density of cohesion energy) of a solvent is to that of a fiber, the better is its swelling or dissolving effect on that fiber. The solubility parameters of aliphatic and aromatic hydrocarbons and their chlorine-substituted derivatives are less than 42 $J^{0.5}cm^{-1.5}$. They are, consequently, proper swelling agents for cellulose triacetate, poly(tetrafluoroethylene), polyethylene, and polypropylene. All the other fibers might be swollen by solvents with solubility parameters exceeding the above mentioned value, for instance water, dimethylsulfoxide, and dimethylformamide. The simplicity of this correlation might be disturbed by parameters such as the crystallinity of the polymer, the temperature of the system, secondary bonds between the fiber-forming polymer and the solvent, and the conditions of fiber formation [16].

5.2 Accessibility, Reactivity, and Stabilization of the Swollen State of Fibers

Most of the molecules remain inaccessible in surface reactions. The proportion of unreacted molecules after microheterogeneous reactions is equal to the proportion of the regions with high lateral order. In the case of macroheterogeneous reactions, an increase in the number of reacting molecules occurs after long contact time. All the molecules participate equally in permutoid reactions. Reactions starting at the surface of the fibers and continuing toward their internal structure are controlled by the diffusion into the fibers.

Studies on processes and reactions preceded by penetration and depending upon the accessibility of fibers indicate that special attention should be paid to the surface of crystallites or microfibrils as well as to changes which precede and enable water penetration into formerly closed areas. Sorption of water vapor by polar fibers, acid hydrolysis of cellulose, and esterification of cellulose can be mentioned as examples.

Changes in fiber structure brought about by swelling can be reversed by drying, unless a treatment to stabilize the intermediate swollen structure has been applied. The swelling of cellulose which occurs during the course of mercerization essentially disappears during the subsequent neutralization, washing, drying, and heat-setting operations. If the water-rinsed, mercerized cotton is subsequently solvent-exchanged with acetone, benzene, and heptane, its swollen state can be retained even after drying in vacuo. Rearrangement of H bonds stabilizes the otherwise thermodynamically labile system. This xerogel-like structure is stable only until its first wetting by water. However, the temporary existence of the unstable structure offers a number of possibilities for new dyeing and finishing processes utilizing the increased reactivity of the cellulose.

The solution formed by the polymer molecules of the fiber and by the swelling liquor can be considered as highly concentrated; thus a strong interaction occurs between the molecules of the two components, which significantly decreases their reactivity toward any third reactant. A remarkable loss in dissolving capacity of the swelling liquor toward a third compound can be accompanied by an increased solubility of that compound in the swollen-fiber phase. Accessibility of a fiber to a compound dissolved in the swelling agent cannot be forecasted according to our present knowledge. This can be demonstrated by a few deviations from the order of reactivity generally accepted as viscose rayon > mercerized cotton > cotton. The adsorbed amount of benzene vapor is higher on cotton than on viscose rayon, and the accessibility of cotton to a solvent mixture containing glacial acetic acid exceeds that of mercerized cotton. Also, the amount of water combined with cellulose from an aqueous sodium thiosulfate solution can be determined by Schreinemaker's method [17]. Cellulose, after having been saturated with the solution, retains water; thus, the concentration of the retained solution expressed from the cellulose exceeds that of the saturating liquor, since water combined with cellulose loses its dissolving capacity toward sodium thiosulfate. The amount of water combined in this manner by 1 mol of unmercerized and mercerized cotton cellulose is 0.5 and 1.0 mol, respectively. The amount of water combined in mercerized cellulose is reduced by drying from 1.0 to 0.7 mol per mol of cellulose.

The rate of cellulose penetration by nonionic organic compounds (e.g., urea, glycerol, sugar) in aqueous solution exceeds that of ionic inorganic compounds of similar molecular weights (e.g., sodium chloride, sodium carbonate, aluminum chloride). An increase in the molecular weight of the penetrating compound brings about a decrease in the rate of penetration. A molecular weight of 5000 can be considered as the upper limit to enable any fiber penetration [18]. The rate of cellulose penetration is significantly decreased by prior crosslinking of the fiber.

Only a small proportion of the solvent in a padding solution can be considered as a swelling agent for the fiber; the rest is simply filling the pores among fibers and yarns. A major proportion of the dissolved compound remains in the pores outside the fibers. Consequently, fiber penetration immediately after the fabric has left the squeeze rolls is still far from equilibrium. This stage can be approached, however, during a properly controlled drying process subsequent to padding. Evaporation of solvent from the interstices outside the fibers is the force compelling the dissolved molecules to penetrate the fibers. Penetration during drying is the opposite of the penetration which occurs during exhaustion dyeing; concentration of the liquid around the fibers increases with time in the former process but decreases with time in the latter one.

Fiber penetration of a solute in the course of solvent evaporation has not yet been explained completely. The solvent molecules leave the surface of the fiber after having arrived at the surface by means of a molecular diffusion mechanism. If the rate of evaporation exceeds that of the diffusion, the border line between dry and wet polymer ranges moves steadily deeper toward the center of the fiber. Diffusion of the solvent from the center of a fiber toward its surface brings about a concentration gradient in the fiber-forming polymer. This generates a concentration gradient of the solute in the opposite direction, promoting the migration of solute toward the center of the fibers [19].

Since diffusion is the rate-controlling unit process for penetration during drying, the time needed to complete the penetration is proportional to the square of the fabric thickness. Only a few seconds are required for complete fiber penetration during drying, while 0.1 μm penetration requires only a few hundredths of a second, even toward the end of solvent evaporation.

In certain cases (e.g., application of the reducing agent in a printing process), the dissolved chemical should be retained in the solution on the surface of the fibers, rather than penetrating them. Due to preferential binding forces, hydrophilic polymers, such as those used as thickeners, can be deposited at the fiber surface to prevent penetration; they do this by promoting an uneven distribution of water and solute between polymer and fabric.

5.3 Prewetting Followed by Penetration of Fabrics

A solute in aqueous solution needs hours or days to diffuse into wet fabrics. The same results can be arrived at in minutes if the solution is circulated through the fabrics by agitating the liquor and fabric at the same time. The water initially filling the pores and cavities of the fabrics has to be exchanged completely by the solution. Preferential adsorption of the solute on the fabric occurs under similar conditions.

Consequences of the First Contact Between Fabric and Liquid: The Thermotex Principle

Although the first contact between fabric and liquid occurs within a fraction of a second, it plays one of the main roles in continuous textile chemical processes [20]. Fabric entering a liquid is represented by a simplified model in Fig. 2.7a; the liquid penetrates the pores and cavities of the fabric while the fabric is moving into the liquid.

The air replaced by the liquid can leave the system only as long as its way is free. One pore of the fabric is represented by a branched capillary in Fig. 2.7b. The departing air is under the action only of hydrostatic pressure. If, however, the air cannot leave the cavities and pores rapidly, bubbles are retained in them (Fig. 2.7c). The bubbles are then kept in motion by buoyancy instead of hydrostatic pressure. In capillary cavities, a bubble is enclosed between two menisci (Fig. 2.8) and the motion of the bubble is influenced by the two contact angles θ_1 and θ_2. Due to the increasing meniscus in the direction of motion, the contact angle decreases on that side and the liquid retreats, thereby breaking the motion of the bubble.

It can be concluded that if the air cannot leave the pores of the fabric immediately after the fabric has entered the liquid, its departure later will be much more difficult. The wetting of the fabric will therefore be determined by the physical conditions characteristic of the system at the moment the fabric enters the liquid. The work needed

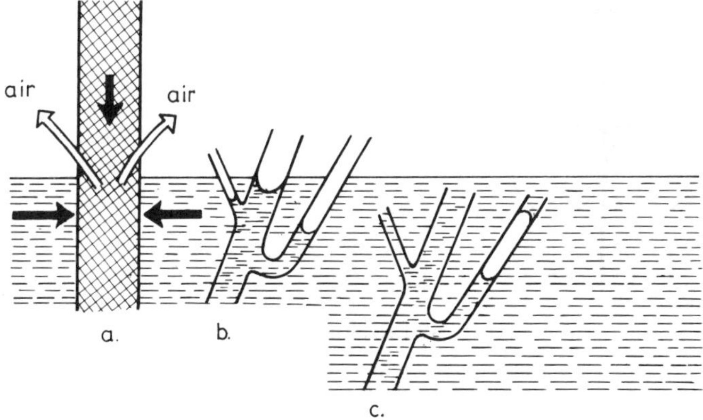

Figure 2.7 Model of a textile entering a liquid: (a) expulsion of air during immersion; (b) and (c) models of porous substrates in the course of immersion.

to expel the air from the fabric can be reduced by inclusion of a wetting agent in the liquid. The same result can be obtained by squeezing out the air occluded in the fabric immediately before saturating it with the liquid; this can be accomplished on equipment such as the padders shown in Fig. 2.9.

Another solution to the problem has been elaborated by Fox et al. [21]; the solution is to remove the air by application of a vacuum. This principle has been applied in padder construction to raise the speed and improve the uniformity of penetration.

The rate and uniformity of textile chemical processes depend upon the complex structure of the fibrous substrate as well as upon the accessibility and partial inactivation of its active sites. Air and water are always present within the complex structure of the fibrous sub-

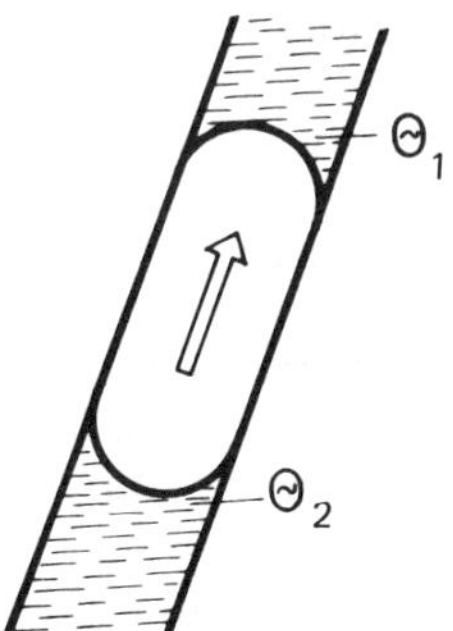

Figure 2.8 Bubble in a capillary.

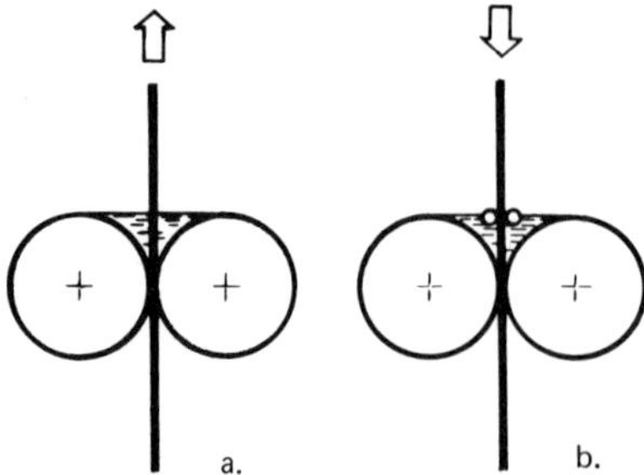

Figure 2.9 Pads improving wetting of fabrics. (a)

strate. The water may be part of the primary or secondary structure of the polymer system; alternatively, it may be adsorbed by the fiber, it may be condensed within the fibers micropores or microcapillaries, or it may simply be filling the macropores or macrocapillaries. The amount and distribution of water bound chemically to the fiber-forming polymer are influential in determining the rate and uniformity of textile chemical reactions. Chemical reactions with the fiber are slowed by water bound to the active sites of the fabric.

When air is filling the pores, wetting of the fibrous substrate is difficult, due to the relatively high water-air interfacial tension and the expansion of the air caused by heat transfer from hot, saturated liquors. The formation of microbubbles, streaming out from the pores as soon as the fibrous substrate enters the hot liquor, is due to the expansion of the air. This airstream counteracts the penetration of liquors into the pores.

The effectiveness and rate of wetting are greatly influenced by the absolute and relative thermal condition of the fiber and liquid. The temperature of dyeing or finishing solutions is very often around 100°C in the course of traditional continuous textile chemical processes, while the fabric is usually at room temperature. Expansion of air included in the pores of the fabric prevents rapid and uniform penetration when cold fabric enters a hot liquor.

Preheating the fabric immediately before it enters the liquid is one of the best solutions to wetting and heat-transfer problems during continuous textile chemical operations. The optimum preheating temperature ranges are: cellulosics, 120-220°C; wool, 105-150°C; synthetics, 10-15°C below the lower limit of their melting range. The higher the temperature of the fabric when entering the liquor, the more reactive are the sites vacated by structurally bound water during preheating. The optimum duration of preheating enables complete removal of structural water without allowing inactivating interactions between neighboring active sites.

When the fabric enters the liquor immediately after preheating, water vapor filling the pores will be condensed and occluded hot air will be contracted. Due to both of these changes, a vacuum will be generated in the pores and capillaries of the fibrous system. Condensation and contraction occur even in liquors at 90-95°C, assuming a sufficiently high temperature of preheating.

Vacuum generation is only the byproduct of the Thermotex process. The main effect is the thermal activation of reactive sites and their immediate reaction with dye or finishing-chemical molecules dissolved in the liquor [22]. The thermal capacity of the liquor is much higher than that of the fabric; therefore, very rapid heat transfer can be expected from liquor to fabric. However, fabric temperature plays an extremely important role at the first moment of contact between liquor and fabric, especially with regard to processes in the pores. The following equation can be used to predict the various effects [17]:

$$\frac{\Delta t}{t_l - t_f} = \frac{c_f G_f}{c_l G_l + c_f G_f}$$

where data pertaining to the fabric substrate are

t_f = temperature

c_f = specific heat

G_f = mass

and data pertaining to the liquor penetrating into the fabric are as defined above except indicated by the subscript l.

Δt = the difference between the temperature of the liquor and that of the solution already penetrated into the pores

The mass and heat capacity of the air filling the pores and holes of the fabric can be neglected, since these values are below those of the fabric and liquor by three orders of magnitude.

For the sake of simplicity, G_f should be taken as 1000 g. The proportion of liquor penetrating the fabric from the total amount present in the fibrous system at the moment of immersion in the batch is only 40-60%. The remaining amount of the liquid is only adhered to the fabric, although some part of it may enter the pores under the squeezing effect of the pad. Assuming a cotton fabric (c_f = 1.26 kJ/kg°C), an aqueous solution (c_l = 4.20 kJ/kg°C) and the pores to be 50% of the weight of the fabric (G_l = 500 g), then

$$\frac{\Delta t}{t_l - t_f} = 0.38$$

If the temperature of the bath is 90°C and that of the fabric at the moment of its entering the liquid is 20°C, then

$$\Delta t = 0.38(90 - 20) = 27°C$$

The temperature of the liquid within the pores will consequently be

$$90 - 27 = 63°C$$

Although the above model assumes uniform heat transfer with very thin capillary (pore) walls and microscopic penetration distances, in reality it is unlikely that the temperature distribution will be uniform within the pores and voids of the fabric. Due to laminar flow of the liquid in the capillaries, heat will be conducted within this liquid. The temperature at the top of the advancing liquid column will be equal to t_f, being constantly in contact with the capillary walls of the temperature (Fig. 2.10). If the fabric is preheated to over 100°C, that part of the liquid column starts boiling and the generated vapor expels air from the pores and voids. This is an easy and rapid process because the flow resistance of air is only 1% that of aqueous solutions. The major part of the energy consumed by preheating of the fabric helps maintain the desired temperatures of the dyeing or finishing operation.

Water vapor, replacing air in the pores and voids of the fabric, will be condensed to water in the moment of entering the bath because the temperature of the bath is always under the boiling point of the water. The condensation of vapor brings about a vacuum within the pores of the fabric which aids the penetration of the finishing bath into the fabric.

Preheating the fabric below 100°C can also prove to be advantageous. For instance, in Fig. 2.11 the length of a capillary of constant diameter is designated as l and the average rate of flow of the liquid in the capillary as W. A column of a mixture of vapor and air of length l-x is pushed forward by the advancing liquid meniscus a distance represented by x. Flow resistance of the vapor-air mixture relative to that of the liquid can be neglected. The difference in pressure between the two ends of the capillary can be calculated with the help of the Poiseuille equation:

$$\frac{8\eta_l x}{r_m^2} w = \frac{2\gamma_{lv} \cos \theta}{r_m} + H\rho_l$$

where

η_l = the viscosity of the liquid

γ_{lv} = surface tension at the liquid-air interface

θ = contact angle

r_m = average diameter of the capillary

H = height of the liquid column controlling hydrostatic pressure in the capillary

ρ_l = density of the liquid

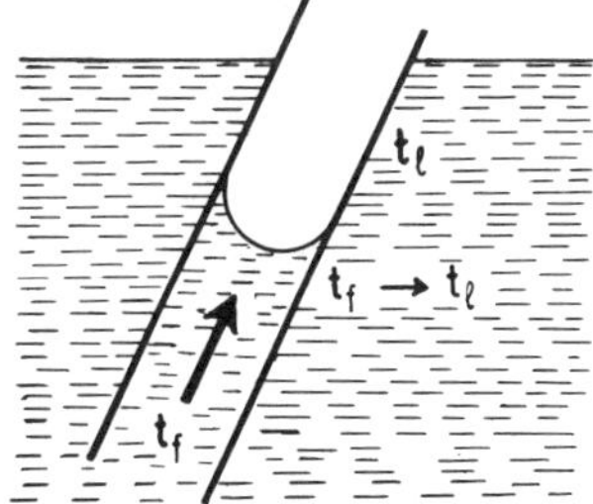

Figure 2.10 Temperatures in and around the capillary.

The term

$$\frac{2\gamma_{lv} \cos\theta}{r_m}$$

is the difference in pressure brought about by the capillary forces. Its value is negative if the contact angle exceeds 90° and can, under certain conditions, prevent the penetration of the liquid. The term

$$\frac{8\eta_l x}{r_m^2}$$

expresses the effect of the viscosity of the liquid on fabric wetting.

The effect of temperature on the viscosity of water is shown in Fig. 2.12. Since the viscosity of water at 100°C is only one-third of the value at 20°C, the flow resistance will decrease with an increase in temperature.

The Poiseuille equation is valid only under isothermal conditions. In the case of cold fabric entering a hot liquor, the thermal effect occurring on the capillary walls influences the amount of liquid which is in laminar flow near those walls by bringing about an increase in liquid viscosity and flow resistance in that region. On the other hand, if at the beginning of the process the temperature of the fabric is

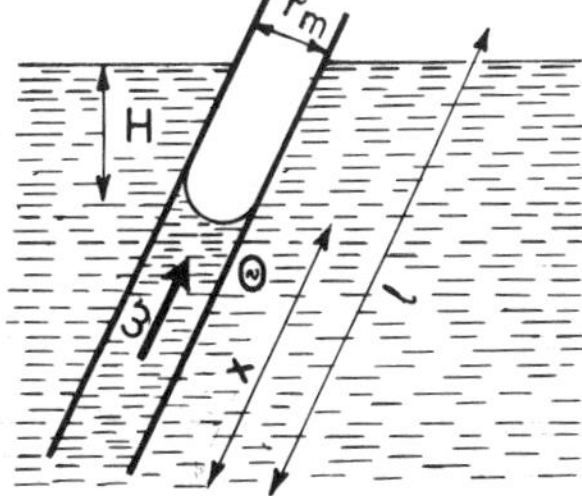

Figure 2.11 Parameters influencing the pressure in a capillary.

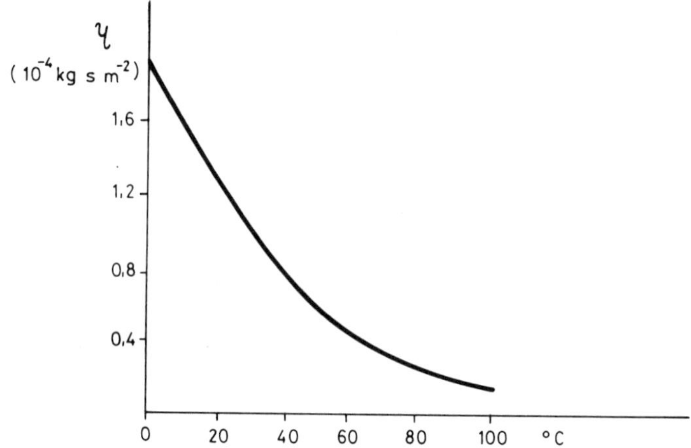

Figure 2.12 Temperature dependence of the viscosity of water.

higher than that of the hot bath, the liquid viscosity, especially next to the meniscus of the column, is significantly decreased.

Fig. 2.13 can be considered as evidence for the increase in the absorbency of a fabric with an increase in its temperature. Factors acting to inhibit absorbency within the pores of fabrics, especially in gray fabric, can be easily overcome by raising the temperature of the fibrous system (Fig. 2.14). Surface tension and pH of the liquid also play important roles in the wetting and saturating processes (Figs. 2.15 and 2.16).

From the measurable apparent density values of fibrous systems, the volume of air retained in the pores of the fibers, yarns, and fabrics

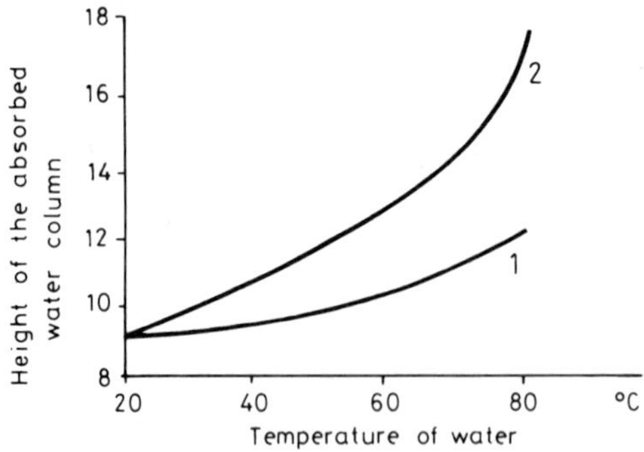

Figure 2.13 Temperature dependence of the height of a water column absorbed by the fabric: (1) The temperature of the fabric is constant at 20°C. (2) The temperature of the fabric is equal to the temperature of the water.

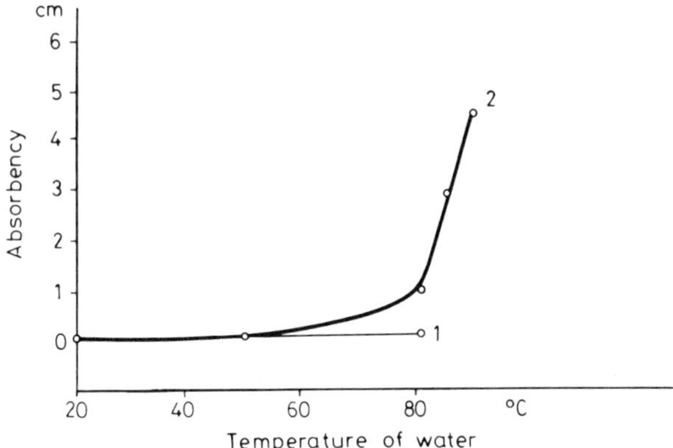

Figure 2.14 Effect of temperature on the absorbency of gray cotton fabric: (1) The temperature of the fabric is constant at 20°C. (2) The temperature of the fabric is equal to the temperature of the water.

can be calculated. If the apparent density is equal to the real density of the fiber, known from the literature, then there is no air included in the pores of the system. The effect of heating up the fibrous system on its apparent density, and thus on its included air content, is shown in Figs. 2.17 and 2.18.

The Effect of Padding on Liquid Pickup of Fabrics

In the course of conventional immersion operations, liquid penetrates only to a limited extent into the fabrics. Squeezing improves penetration by pressing liquid from the surface of the fabric and from its large pores into smaller pores. Complete saturation, however, cannot be arrived at in this way but requires more effective methods such as

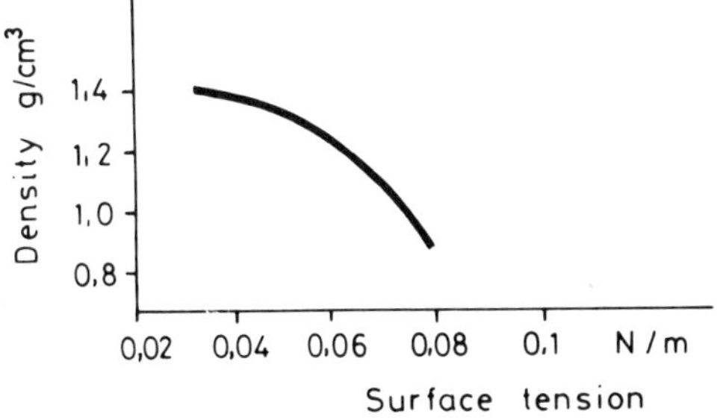

Figure 2.15 Dependence of the apparent fiber density on the surface tension of the saturating liquid.

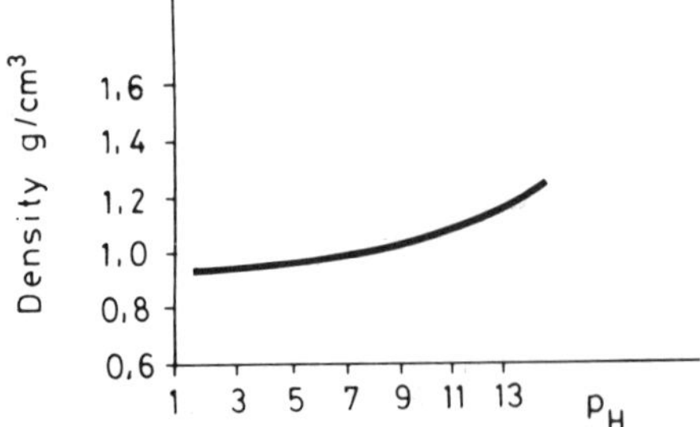

Figure 2.16 Dependence of the apparent fiber density on the pH of the saturating liquid.

the application of vacuum or the preheating of the fabric in combination with padding. Table 2.8 shows how vacuum padding improves liquid penetration.

Comparing the data of Table 2.8 with those of Figs. 2.16 and 2.17, it can be concluded that to reach an apparent density of 1.42 on gray cotton, the fabric must be preheated to at least 90°C.

Relationships between pickup values and the chemical nature of the liquid are shown in Table 2.9. From this data we can conclude the following:

1. The same volume of liquid is retained in the fabric upon squeezing under the same nip.
2. Caustic solutions, due to their intrafibrillar swelling effect on cotton cellulose, cannot be squeezed out as easily as water.
3. Temperature and concentration of the caustic solution have an influence on changes in the amount of liquid retained by the fabric after squeezing similar to that they have on changes in the degree of intrafibrillar swelling.

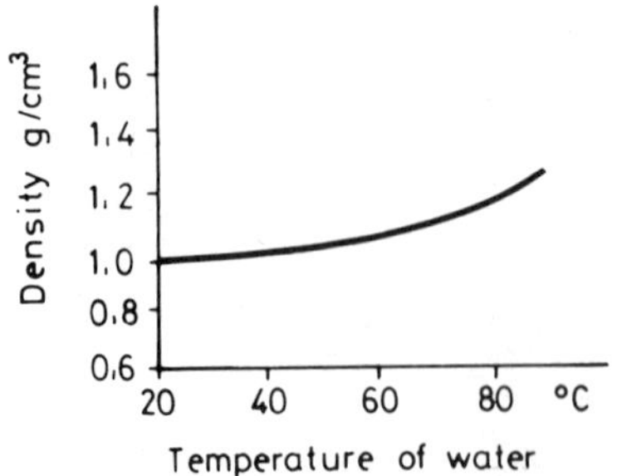

Figure 2.17 Dependence of the apparent density of gray cotton fabric on the temperature of the saturating water.

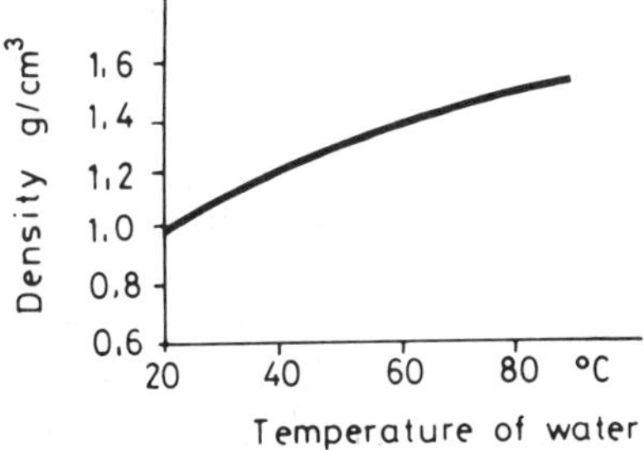

Figure 2.18 Dependence of the apparent density of gray cotton fabric, preheated to the temperature of the saturating water, on the temperature of the latter.

5.4 Thermokinetics of Liquid Penetration into Cellulose

The availability of the microcalorimeter of Calvet [23], used earlier for the determination of the heat of immersion, also enables the thermokinetic study of liquid penetration into fibers, such as that carried out by Balczerzyk and co-workers [24,25]. The thermokinetics of a

Table 2.8 Effect of Squeezing on Water Pickup at 20°C

Operation	Nip (bar)	Apparent density (g/cm^3) of cotton fabric: Gray	Scoured
Immersion into water for 1 min	0	0.60-0.70	1.30-1.35
Immersion into water followed by squeezing	15	0.65	1.43
Immersion into water followed by squeezing	44	0.51	1.51
Immersion into water between two squeezing operations	15	0.78	1.45
Immersion into water between two squeezing operations	44	0.78	1.48
Vacuum padding	15	1.42	1.52
Vacuum padding	44	1.42	1.52

Source: Adapted from Ref. 21.

Table 2.9 Liquid Pickup of Saturated and Squeezed Scoured Cotton Fabric

	Liquid characteristics temperature	density	Pickup	
Liquid	(°C)	(g/cm^3)	(g)	(cm^3)
Distilled water	20	0.998	99.1	99.5
Kerosene	20	0.824	82.5	100.7
Spindle oil	20	0.915	88.3	96.5
Sodium hydroxide solution, 4%	20	1.045	103.0	98.5
Sodium hydroxide solution, 20%	20	1.190	140.8	118.0
Sodium hydroxide solution, 20%	90	1.170	114.2	97.6

Source: Adapted from Ref. 17.

process are represented by a function in which the rate of heat release is plotted against time.

In the course of the wetting of cellulose, hydrogen bonding occurs between hydroxyl groups, ring and glycosidic oxygen atoms of cellulose, and the corresponding atoms in water, alcohols, or acetic acid. These solvents penetrate only the accessible regions of cellulose. Heat release following penetration can be used as the relative measure of accessibility.

The mechanism of penetration described by the thermokinetics of the process is highly dependent upon the structure of the accessible regions. Description is possible with the modified equation of diffusion [26].

It is well known that the rate and extent of penetration into cellulose, in the case of nonreactive liquids, depends upon the dimensions of the liquid molecules as well as on the structure of the accessible regions of cellulose. Aqueous wetting of cellulose, due to its very high rate, is not a suitable method for accurate thermokinetic studies of cellulosic fibers of various origins or treatments. In the case of native cellulose, ethanol and acetic acid are two solvents with the proper molecular size to penetrate at a sufficiently slow rate to enable the determination of differences in the structure of the accessible regions. Methanol is a similar solvent for regenerated cellulose. Penetration by solvents of smaller molecular size is too fast. Larger molecules (e.g., propanol) cause thermograms with more than one peak, indicating discontinuous distribution of the accessible regions of the cellulose.

Native and regenerated cellulose can be distinguished by aqueous wetting, since the integral heat of wetting of the latter is 2.5 times higher than that of the former. The rate of wetting of cellulose by methanol is as rapid as its wetting by water. Interestingly, the process with methanol is only slightly slower on mercerized native cellulose, but significantly slower on regenerated cellulose than on native cellulose. The integral heat of wetting with methanol is less than with water.

The rate of heat release is definitely slower in the course of wetting with ethanol than with water or methanol, particularly in the case of native cellulose. Two kinetically different sections of penetration can be distinguished on the thermograms. The first is a rapid section with a well-defined peak, while the second is a very slow, characteristically flattened section.

Using propanol or n-butanol to wet native cellulose, the rate of heat release is faster than with ethanol; the integral heat of wetting, however, keeps decreasing. No change in thermokinetics occurs with regenerated cellulose.

Although the size of glacial acetic acid molecules is nearly equal to the size of ethanol molecules, the integral heat of wetting on native cellulose is twice as great with the former than with ethanol. The rate of heat release is, however, three times slower with glacial acetic acid. No thermokinetic difference is detected on regenerated cellulose when comparing the effects of the two solvents.

To detect changes in the fine structure of native cellulose with thermokinetics, the use of ethanol or glacial acetic acid has proved to be advantageous. The use of methanol is suitable with regenerated cellulose.

Preheating dry cellulose to 105°C brings about significant changes in the structure of its accessible regions. The degree of change varies with the type of cellulose. A noticeable change occurs on the thermokinetic plot of native cellulose. With mercerized native cellulose or regenerated cellulose, the rate of heat release decreases to one-third of that of the untreated cellulose samples.

The change in accessibility brought about by preheating is reversible. By drying the sample in vacuo and rewetting with water, the original values of the integral heat of wetting and the original thermograms can be regained.

An important role is played by the medium of the last treatment preceding vacuum drying of the fiber. Very slow penetration of ethanol or acetic acid combined with great integral heat of wetting occurs on cotton dried from water, while very rapid penetration combined with definitely smaller integral heat of wetting is the consequence if the cotton has been dried from ethanol. No such difference could be observed, however, on regenerated cellulose under similar conditions. No explanation for this effect is available.

Penetration of sodium hydroxide solutions into cellulose is concentration dependent. Solutions of concentrations higher than 21% penetrate also the highly ordered regions of the fiber, although the rate of their penetration is lower than that of caustic solutions of concentrations below 21%. It could be demonstrated [27] that preferential water adsorption occurs on cellulose from solutions below the 21% concentration limit. The rate of penetration of free water molecules (not bound as solvates) is much higher than that of the large solvated ions of sodium hydroxide. In solutions above the described concentration limit, there are no longer any free water molecules. Rate of penetration is determined by the much slower motion of the complex solvated ions. The process is therefore characterized by a low rate of penetration. A further significant drop in the rate of penetration occurs with caustic solutions above a concentration of 37.5%. The gross integral heat of the wetting process is, however, less at 21% concentration than at 37.5% or higher concentrations.

According to Hearle [28], continuous distribution in the order of molecules is characteristic of the intramicrofibrillar structure. Penetration into intramicrofibrillar regions is controlled by the size of the solvent molecules as well as by the nature of their functional groups. Distances between adjacent structural elements are long enough in the accessible regions of cotton or mercerized cotton microfibrils to enable unhindered penetration of water and methanol. Similar penetration in regenerated cellulose is possible only for water. Solvent molecules of a size corresponding to the average value of the distances within regions of low lateral order, e.g., ethanol or acetic acid for cotton and methanol for regenerated cellulose, cannot move unhindered in the intramicrofibrillar channels. This penetration is, consequently, of a lower rate.

Intermicrofibrillar distances (voids) are larger than the intramicrofibrillar ones. Consequently, solvents even of higher molecular weight penetrate rapidly, without any diffusion control, into intermicrofibrillar regions.

Loss in the integral heat of wetting and slowing down of penetration brought about by thermal pretreatment at 105°C can be explained by assuming that the higher intensity of polymer segment motion at elevated temperatures favors the formation of hydrogen bonds among adjacent fiber-forming molecules. This assumption corresponds also with the reversible nature of the process. As the average hydrogen bonding energy value is higher in cotton than in regenerated cellulose, this can explain the different behavior of the two fibers [29].

The consequences of thermal pretreatment include not only the decrease in rate of methanol penetration but also often the appearance of thermograms with two peaks. Similar thermograms can occur with regenerated cellulose in the course of wetting with caustic solutions of high concentration. Thermograms with two peaks have two different explanations:

1. The first peak is assigned to the rapid solvent penetration into the intermicrofibrillar regions, the second one to the much slower penetration into the intermolecular regions of low lateral order.
2. The concept of zonal structure cannot be neglected in the case of regenerated cellulose. This corresponds with varying densities of molecular packing in adjacent zones. In certain zones penetration of the solvent is rapid (first maximum), while in zones of higher densities it is very slow (second maximum). Supporting this interpretation is the fact that no thermogram with two maxima can be obtained even with caustic solution in the case of superpolynosic fiber of monozonal structure [30].

5.5 Carrier Supported Penetration of Fabrics

The rate of diffusion of disperse dyes into fibers is very low unless special measures are taken. An accelerating effect can be realized by the use of temperatures above 100°C or the addition of carriers such as hydrocarbons, phenols, primary amines, and esters to the dyeing system. Due to numerous actions of carriers, no unambiguous explanation has yet been accepted as to their effect. Their main effect has been assigned by one group of researchers to the increased solubility of the dyestuff, either in the dyeing medium or in the carrier itself [31,32], while others have assigned it to swelling of the fibers and increased segmental motion of the polymer molecules [33].

The significance of the changes in certain physical fiber characteristics brought about by carriers has been pointed out [34]. For instance, shrinkage of synthetic fibers has been observed under the action of effective carriers [35]. Also, a decrease in the glass transition temperature T_g of the fiber-forming polymer could be detected paralleling the increase in carrier pickup (Fig. 2.19).

The increase in length of a fiber brought about by a given load plotted against time (creep curve) is known at temperature T as *strength viscosity* (μ'_T). The μ'_T curves, even of the same fiber, are different at different temperatures. Two curves at different temperatures, plotted logarithmically, can be made to overlap by means of a lateral shifting. The length of the shift required to bring about the overlapping is expressed by the shift factor a_T, the ratio of viscosity values occurring at the two temperatures T and T_g. This shift factor primarily reflects the temperature dependence of polymer segmental friction or mobility, which influences the rate of all configurational rearrangements.

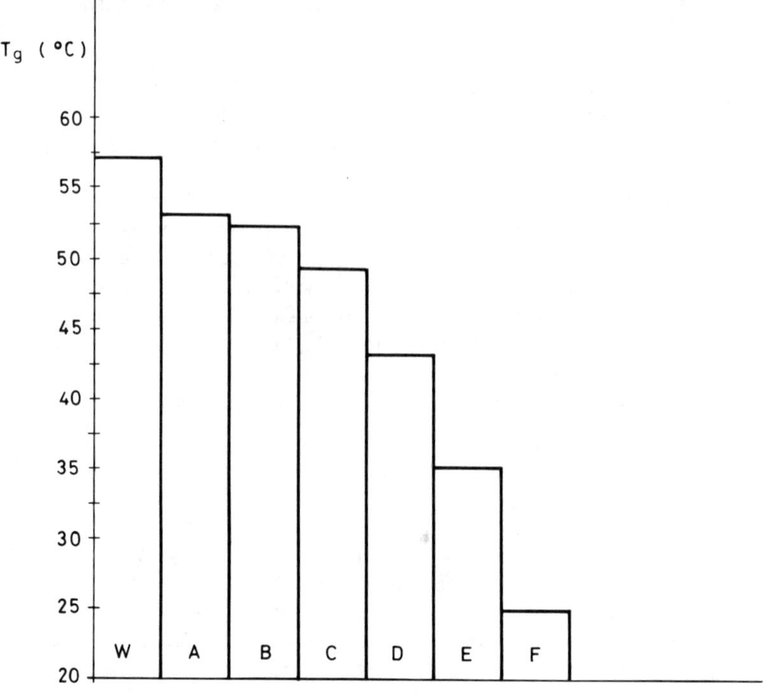

Figure 2.19 Glass transition temperature of an acrylic fiber changing in the presence of different carriers. Adapted from Ref. 36.

Empirical relationships have been derived between the shift factor of the creep curve and the corresponding T_g values; one successful one is expressed by the equation of Williams, Landel, and Ferry [37]:

$$-\ln a_T = \frac{A(T - T_g)}{B + (T - T_g)}$$

where

$$a_T = \frac{\mu_T}{\mu_{T_g}}$$

This Williams-Landel-Ferry equation has been modified for the diffusion of dyes into fibers by Rosenbaum [38]:

$$-\ln a_T + \ln \frac{D_T}{D_{T_g}}$$

Consequently,

$$\ln \frac{D_T}{D_{T_g}} = \frac{A(T - T_g)}{B + (T - T_g)}$$

where D_T and D_{T_g} are the diffusion coefficients at temperatures T and T_g, respectively. These four values can be experimentally determined so that the constants A and B can be calculated for a given fiber system.

It can be concluded that the segment motion of the fiber-forming polymer molecule controls not only the viscoelastic characteristics of the fibers but also the penetration of fibers by the dyestuffs. The effect of a carrier appears to be to plasticize the polymer and, by this means, decrease its glass transition temperature. The above relationship is equally valid for polyester fiber-disperse dye and for polyacrylonitrile fiber-cationic dye systems [39].

REFERENCES

1. E. Valko, *Chemical Aftertreatment of Textiles* (H. Mark, N. S. Wooding, and S. Atlas, Eds.), Wiley Interscience, New York (1971).
2. E. Wolfram, *Kolloidika*, Tankönyvkiadó, Budapest, (1976).
3. W. Bobeth and H. I. Jakobasch, *Melliand Textilber.* 55, 268 (1974).
4. J. W. Rowen and R. L. Blain, *Ind. and Eng. Chem.* 25, 1663 (1947).
5. S. Máttyus, *Magyar Textiltechnika* 30, 342 (1977).
6. B. C. Burdett, *The Theory of Coloration of Textiles* (C. L. Bird and W. S. Boston, Eds.), Dyers, Bradford, Eng. (1975).
7. A. J. Hailwood and S. Horrobin, *Trans. Faraday Soc.* 42B, 84 (1946).
8. J. W. S. Hearle and R. H. Peters, *Moisture in Textiles*, Butterworths, Manchester, England (1960).
9. J. J. Kipling, *Adsorption from Solutions of Non-Electrolytes*, Academic, London and New York (1965).
10. C. L. Bird and F. Manchester. *J. Soc. Dyers Col.* 71, 604 (1955).
11. R. H. Peters, *The Physical Chemistry of Dye Adsorption* (I. D. Rattee and M. Breuer, Eds.), Academic, London (1974).

12. W. McDowell and S. Weingarten, *J. Soc. Dyers Col.* *85*, 595 (1969).
13. S. R. S. Iyer, G. Srimrasan, and N. T. Baddi, *Text. Res. J.* *38*, 693 (1968).
14. R. H. Peters, *The Physical Chemistry of Dyeing*, Elsevier, Amsterdam (1975).
15. H. J. White and H. J. Eyring, *Text. Res. J.* *17*, 523 (1947).
16. P. A. Small, *J. Appl. Chem.* *3*, 71 (1953).
17. I. Rusznák, *Magyar Testiltechnika* *32*, 329 (1979).
18. H. B. Mann and T. H. Morton, *Disc. Faraday Soc.* *16*, 58 (1954).
19. M. S. Aboul-Fetouh and L. W. C. Miles, *Text. Res. J.* *38*, 176 (1968).
20. I. Rusznák, T. Bonkáló, and J. Sármány, *Melliand Textilber.* *54*, 1087 (1973).
21. M. R. Fox, W. J. Marshall, and N. D. Stewart, *J. Soc. Dyers Col.* *83*, 493 (1967).
22. I. Rusznák, T. Bonkáló, R. Beszeda, and I. Bellada, *British Patents 1,038,205* (1963).
23. Calvet, *Zh. Fiz. Khim.* *33*, 1161 (1959).
24. E. Balcerzyk and W. Kozlowski, *Przeglad Papierniczy* *20*, 309 (1964).
25. E. Balczerzyk, W. Kozlowski, and G. Wlodarski, *Text. Res. J.* *39*, 666 (1969).
26. J. Crank, *The Mathematics of Diffusion*, Clarendon, Oxford (1957).
27. A. I. Kapustinski and O. J. Samoilov, *Zh. Fiz. Khim.* *26*, 918 (1952).
28. J. W. S. Hearle and R. H. Peters, *Fibre Structure*, Butterworths, London (1963).
29. K. J. Heritage, J. Mann, and L. Roldán-González, *J. Polym. Sci.* *A1*, 671 (1963).
30. G. W. Nikonovich, S. A. Leontieva, H. D. Byrkhanova, and K. V. Usmanov, *Khim. Volokna* *5*, 54 (1965).
31. C. L. Zimmerman, J. M. Mecco, and A. J. Carlino, *Amer. Dyestuff Reptr.* *44*, 296 (1955).
32. T. Vickerstaff, *Hexagon Digest* *20* (7) (1954).
33. V. S. Salvin, *Amer. Dyestuff Reptr.* *49*, 600 (1960).
34. A. Murray and R. Mortimer, *J. Soc. Dyers Col.* *67*, 173 (1971).
35. F. M. Rawicz, D. M. Cates and H. A. Rutherford, *Amer. Dyestuff Reptr.* *50*, 320 (1961).
36. R. Fujino, W. Kuroda, and F. Fujimoto, *J. Soc. Fibre Sci. Tech. Japan* *21*, 573 (1965).
37. M. L. Williams, R. F. Landel, and J. D. Ferry, *J. Amer. Chem. Soc.* *77*, 3701 (1955).
38. S. Rosenbaum, *J. Appl. Polym. Sci.* *9*, 2071; *9A*, 2085 (1965).
39. W. Ingamells, R. H. Peters, and S. E. Thornton, *J. Appl. Polym. Sci.* *17*, 3733 (1973).

3

ALKALI TREATMENT OF CELLULOSE FIBERS

RENÉ FREYTAG École National Supérieure de Chimie de Mulhouse, Mulhouse, France

JEAN-JACQUES DONZÉ Centre de Recherches Textiles de Mulhouse, Mulhouse, France

1. Introduction 94

2. Action of Alkaline Agents on Cellulose Fibers 94

   2.1 Hydration of alkali metal hydroxyde ions 94
   2.2 Mechanism of alkali cellulose formation 96
   2.3 Swelling and solubility of cellulose fibers in alkali solutions 100
   2.4 Degradation of cellulose in alkali solutions 107

3. Scouring of Cotton 111

   3.1 Introduction 111
   3.2 Impurities of cellulosic materials 112
   3.3 Physicochemical aspects of the scouring process 120
   3.4 Technological aspects of the scouring process 126
   3.5 Special scouring processes 132

4. Mercerization of Cotton Fibers 134

   4.1 Introduction 134
   4.2 Influence of mercerization on the structure of cellulose fibers 135
   4.3 Influence of the mercerization process on the properties of cotton fibers 140
   4.4 Auxiliary products for mercerization: wetting agents 145
   4.5 Hot mercerization process 146
   4.6 Mercerization of blend fabrics 148
   4.7 Technological aspects of the mercerization process 151

References 157

1. INTRODUCTION

Among the various pretreatments of cellulosic textile materials, only scouring and mercerization employ an alkaline agent in concentrated solution. In other processes the alkaline agent is in a diluted solution and is employed exclusively for the purpose of obtaining a pH value that will permit either a better solubilization of certain impurities or a reaction mechanism necessary for the purification of fibers (e.g., desizing with oxidizing agents or removing soluble products, bleaching with sodium hypochlorite or hydrogen peroxide).

Although these two alkaline processes, namely scouring and mercerization, use the same reagent, they have different objectives. The scouring or boiling-off process permits the removal of certain impurities with which the fiber is associated. (Impurities are easily removed by alkaline boiling followed by bleaching.) In fact, one of the main advantages of cotton lies in its resistance to alkali solutions; this property enables the cleansing of the raw material in a manner that is not possible with wool or silk. The scouring process, while purifying the α-cellulose, imparts the hydrophilic character and permeability necessary for the subsequent processes (bleaching, mercerizing, dyeing, or printing).

Mercerization is not a fiber purification process, but it induces desirable changes in cotton yarn and fabric properties, e.g., mechanical properties, dyeing properties, and luster.

Before analyzing in detail these two alkaline processes, it is advisable to refer to the interactions between alkaline and cellulosic substances, and first, to the nature of the hydroxides of alkali metals in solution.

2. ACTION OF ALKALINE AGENTS ON CELLULOSE FIBERS

2.1 Hydration of Alkali Metal Hydroxide Ions

There is evidence of some relationship between the degree of hydration of alkali hydroxide ions and their ability to penetrate into cellulose fibers, and consequently their swelling or nonswelling action; hence, the size of the alkali hydroxide hydrates in regard to alkali treatments of cellulose fibers, especially mercerization, is of great importance. The various hydrated (or solvated) forms of the alkali hydrate ions (with correspondence to their sizes) are listed in Table 3.1 [1].

For small concentrations of electrolyte (for example, sodium hydroxide), we notice the presence of pairs of separated hydrated ions. The diameter of these hydrated ions is too large for easy penetration into the macromolecular structure of cotton. As the concentration increases, the number of water molecules available for the formation of hydrates decreases and hydrated ion pairs, dipole hydrates (solvated

Table 3.1 Solvated Forms of Alkali Hydroxide Ions

Type of hydrate	Hydrodynamic diameter (Å)
Dipole hydrate	5-8
Solvated dipole hydrate	8-10
Hydrated ion pair	10-15
Separately hydrated ions	15-20

Table 3.2 Sodium Hydroxide Hydrates: Composition, Type, and Concentration of Maximum Frequency

Composition of hydrate	NaOH concentration (percent by weight)	Type of hydrate	References
NaOH. $20H_2O$	6-9	Hydrated ion pair	2
NaOH. $12H_2O$	13.5-15	Hydrated ion pair	2, 3, 4
NaOH. $10H_2O$	18	Solvated dipole hydrate	2, 4
NaOH. $7H_2O$	22.8-24.1	Solvated dipole hydrate or dipole hydrate (diameter 7.4Å)	4, 5
NaOH. $5H_2O$	30.2-30.9	Dipole hydrate	4, 5, 7, 8
NaOH. $4H_2O$	34.8-35 35.7	Dipole hydrate (α and β forms)	4, 7 5
NaOH. $3.5H_2O$	38-38.8	Dipole hydrate	5, 6, 7
NaOH. $3.1H_2O$	41.8-42.6	Dipole hydrate (diameter 6 Å)	5, 6, 7
NaOH. $2H_20$	52.6	Dipole hydrate	
NaOH. H_2O	69	Dipole hydrate	5, 6

Table 3.3 Potassium Hydroxide Hydrates: Composition, Type, and Concentration of Maximum Frequency

Composition of hydrates	KOH concentration (percent by weight)	Type of hydrate	References
KOH. $24H_2O$	7	Hydrated ion pair	1
KOH. $8H_2O$	25-30	Hydrated ion pair	1
KOH. $4H_2O$	43.8-47	Dipole hydrate	1, 5, 8, 9
KOH. $2H_2O$	55-60.9	Dipole hydrate	1, 5, 7
KOH. H_2O	75.7	Dipole hydrate	5

or not) are formed with decreasing hydrodynamic diameter. Dipole hydrates, because of their small diameter are capable of penetrating not only into the amorphous parts of cotton fibers but also into the parts with higher lateral order, simultaneously forming hydrogen bonds with the molecular chains of cellulose.

The degree of hydration of isolated ions, ion pairs, or dipoles can be assessed by the crystallization of the alkali hydroxide solutions. The concentrations in which the various hydrates are found in the greatest proportions are summarized in Tables 3.2 (for sodium hydroxide) and 3.3 (for potassium hydroxide).

In the case of lithium hydroxide, we can count three hydrates: a dipole hydrate $LiOH \cdot H_2O$ for 58% LiOH by weight [1,5]; a solvated dipole hydrate $LiOH \cdot 10\text{-}12H_2O$ for 9-11% LiOH by weight [1,2]; and finally a hydrated ion pair $LiOH \cdot 17$ or $20H_2O$ for 5-6% LiOH by weight [1-3].

It appears that the level of the penetration of the alkalis and their hydrating water molecules into the fine structure of the cellulose fibers depends on the concentration of the alkali hydroxide employed. Thus, mercerization induces important structural modifications due to the presence of smaller hydrates in the highly concentrated alkali hydroxide solution.

2.2 Mechanism of Alkali Cellulose Formation

In the presence of less concentrated sodium hydroxide solutions various ternary complexes can form between cellulose, sodium hydroxide, and water. In the complexes called soda celluloses (Na-Cell) a certain number of hydrating water molecules of the sodium hydroxide hydrates are replaced by hydroxyl groups of cellulose. Soda celluloses of a crystalline character are well defined according to their x-ray diffraction diagrams. According to Sobue, Kiessig, and Hess [10], the formation of each type of soda cellulose occurs at well-defined NaOH concentrations and temperatures, as seen in Fig. 3.1.

Soda cellulose I is formed by treatment of cotton or ramie at 20°C with sodium hydroxide concentrations of 12-19% by weight at, or about 130-230 g/liter [1,2,7,11,12]. The decomposition of soda cellulose I occurs at the start of the cellulose II formation.

Soda cellulose II seems to form at sodium hydroxide concentrations varying between 20 and 45% by weight [1,7,11]. However, the existence of soda cellulose II has been shown already in 6% (by weight) sodium hydroxide solutions, which would prove that definite compounds of soda cellulose do not necessarily occur in well-defined concentration ranges [13].

Soda cellulose III is obtained by the drying of soda cellulose I [1,13]. By rewetting soda cellulose II, soda cellulose I is formed again [13]. For temperatures higher than 30°C and in solutions of

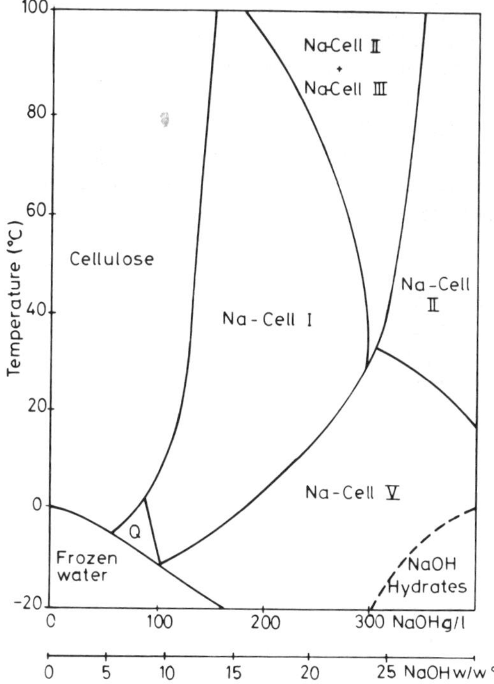

Figure 3.1 Phase diagram of soda celluloses. (*Source*: Ref. 10).

20-25% by weight of sodium hydroxide, we can observe the formation of soda cellulose III at the same time as soda cellulose II [10].

Soda cellulose IV is obtained by washing soda celluloses I and II in water or in a diluted solution of sodium hydroxide [7,11,14]. In fact, this washing induces the decomposition of ternary complexes because the compound formed is only a cellulose hydrate free of alkali [11], thus its name, water-cellulose. Indeed, the x-ray diffraction diagram of this compound subsists after a complete washing of the cellulose. This intermediate product plays an important role during the rinsing after mercerization.

Soda cellulose V appears between -10 and +20°C in a wide range of sodium hydroxide concentrations (12-28% by weight, about 150-300 g/liter), which are of great interest for the mercerization process [1,7,10,16]. The drying of soda cellulose V leads to the formation of soda cellulose II [16]. The dilution of soda cellulose V, an operation which takes place during rinsing after mercerization, would at first induce the appearance of soda cellulose IV and finally give rise to the formation of cellulose II [10].

The existence of soda cellulose Q (Q from the German word *Quellung*, i.e., "swelling") has been shown by treating ramie in 7-9% by weight solutions of sodium hydroxide between -5 and +1°C [1,10]. In such solutions we observe a high swelling of ramie, the value of which could reach 1000% [7]. Soda cellulose Q is a compound free of alkali because the sodium hydroxide hydrates present, especially $NaOH \cdot 20H_2O$ but also $NaOH \cdot 12H_2O$, which begin to appear in significant proportions in spite of a high degree of swelling, are too large to penetrate the regions with high lateral order. Consequently, the sodium hydroxide molecules cannot combine with crystalline cellulose, but the opening of the fiber texture consecutive to the swelling permits the formation of a cellulose hydrate [1,2]. In fact, soda cellulose Q and soda cellulose IV appear to be the same compound, which implies that soda cellulose IV can be directly formed from cellulose I [10].

At a certain concentration of sodium hydroxide (about 40-45% by weight), soda cellulose compounds are replaced by sodium cellulosate, Cell-ONa, the formation of which has also been observed in small quantities at lower concentrations [17].

Concerning the formation of soda cellulose compounds, the number of hydrated molecules of sodium hydroxide which are replaced by a glucosidic group $C_6H_{10}O_5$ could be 2 [10] or 3 [17]. In fact, there is some approximation regarding the real formulae of soda celluloses as well as the manner in which they form with respect to the nature of sodium hydroxide hydrates present in concentrated solutions. Table 3.4 gives various possibilities as to the identity of soda celluloses. As in the case of sodium hydroxide, similar compounds form with other alkali metal hydrates, such as potassium and lithium hydrates.

Two of the potassium cellulose compounds have been identified: potassium cellulose I (K-Cell I), which is formed with ramie in a

Table 3.4 Soda Celluloses: Composition and Corresponding Sodium Hydroxide Hydrates

Soda cellulose type	Composition	Corresponding hydrate	References
Na-Cell I	$C_6H_{10}O_5$, NaOH, $3H_2O$	NaOH. $5H_2O$	10, 18
		NaOH. $7H_2O$	1, 2, 7
		NaOH. $10H_2O$	2
		NaOH. $12H_2O$	2
	$(C_6H_{10}O_5)_2$, NaOH. $2H_2O$		16
	$(C_6H_{10}O_5)_2$, NaOH. $1.5H_2O$		19
Na-Cell II	$C_6H_{10}O_5$, NaOH. H_2O		18
		NaOH. $3\text{-}3.5H_2O$	10
		NaOH. $2H_2O$	7
Na-Cell III	$C_6H_{10}O_5$, NaOH. $2H_2O$	NaOH. $4H_2O$	7, 10
Na-Cell IV (Q)	$C_6H_{10}O_5$, H_2O		10
	$C_6H_{10}O_5$, 0-0.3NaOH. H_2O	NaOH. $20H_2O$	1
		NaOH. $12H_2O$	7
	$C_6H_{10}O_5$, 1.3 or $1.5H_2O$		20
Na-Cell V	$C_6H_{10}O_5$, NaOH. $4\text{-}5H_2O$	NaOH. $5H_2O$	7
		NaOH. $7H_2O$	10

concentration range between 14 and 25% by weight of potassium hydroxide. At above 25% potassium hydroxide, potassium cellulose II (K-Cell II) appears, which yields potassium cellulose I while drying [1,21]. Like Na-Cell IV, K-Cell I is free from alkali and is similar to a cellulose hydrate. The corresponding potassium hydroxide hydrate is $KOH \cdot 8H_2O$, whereas the hydrate should be $KOH \cdot 4H_2O$ at the start of the formation of K-Cell II [1].

For lithium hydroxide concentrations between 5 and 15%, a lithium cellulose compound (Li-Cell I) has been identified, corresponding to the hydrate $LiOH \cdot 10\text{-}12H_2O$ [1,11,22]. While potassium and lithium celluloses have only limited importance for the practical textile finishes, soda celluloses play a preponderant role during treatments of cellulose fibers in concentrated alkali solutions.

2.3 Swelling and Solubility of Cellulose Fibers in Alkali Solutions

Swelling

It was the English chemist John Mercer who in 1850 first drew attention to the fact that cotton cloth immersed in a sodium hydroxide solution shrinks in width and length; at the same time it becomes thicker and tighter under the swelling action [23].

When cotton is treated with a concentrated alkali solution, the fibers swell together with a deconvolution. Their cross section, which initially has an irregular shape similar to a bean, becomes more and more circular, while the lumen disappears [24,25]. The inner secondary wall swells in a more pronounced way than the primary wall and the cuticle.

The difference in swelling between the mature and immature fibers is not significant in water [25], but it is more so in alkali solutions; the immature fibers swell, in general, less than the normal fibers. This fact is the basis of a method of estimating the maturity of cotton fibers [26].

In the swelling of cotton fibers two distinct processes are to be taken into account: one is the *interfibrillary* and the other is the *intrafibrillary* [4,27]. The interfibrillary swelling takes place in water and weak alkali solutions (sodium hydroxide concentration less than 10-15% by weight). In this case, only the amorphous parts are affected by the swelling phenomenon. In more concentrated alkali solutions, the swelling becomes intrafibrillary with the participation of higher lateral order parts of the fiber. This is consistent with the decreased size of alkali hydroxide hydrates when the concentration increases; thus, for a concentration of sodium hydroxide higher than 10-15% by weight, the hydrate $NaOH \cdot 10H_2O$ appears in a noticeable proportion. In contrast to hydrated ion pairs, this solvated dipole hydrate is able to enter the crystalline parts of cotton [2].

Influence of Alkali Concentration on Swelling. In solutions with an increasing concentration of sodium hydroxide at 20°C, the swelling of cellulose (cotton linters) increases to a maximum value for a concentration equal to 8-9% by weight of sodium hydroxide; next it decreases to a minimum (12-13% NaOH by weight) and slowly increases in higher concentration ranges [7].

More recently, swelling evaluations based on precise measurements of thickness variations of cotton fabrics in alkali solutions have shown that the swelling would first increase moderately up to a sodium hydroxide concentration of 120 g/liter (11% by weight) and then to a much greater extent between 120 and 180 g/liter. Starting at 180 g/liter (15-16% NaOH by weight), the swelling value remains constant up to 300 g/liter (23-24% by weight) [28]. These findings would allow a connection between the effect of the mercerization of cotton and swelling. However, we must consider that the maximum swelling of cotton fiber occurs at a sodium hydroxide concentration of about 150 g/liter (13% by weight), while the maximum mercerization effect takes place only at higher concentrations indicating that the swelling is not the only determining factor.

At these concentrations the sodium hydroxide hydrates present are able to enter the crystalline parts of cellulose which explains the mercerizing effect even for a lower degree of swelling; in fact, in a 150 g/liter of sodium hydroxide solution, the maximum swelling observed corresponds uniquely to the swelling of the amorphous parts which cannot cause the mercerizing effect. Results of numerous studies concerning the influence of sodium hydroxide concentrations on cellulose fiber swelling are presented in Table 3.5 for a number of cellulose fibers.

The fact that, in general, the highest swelling occurs at lower alkali concentrations may be attributed to the greater degree of hydration of the alkali ions in diluted solutions [31].

Influence of Alkali Metal. The other alkali hydroxides behave similarly to sodium hydroxide. Measurements of the increase in thickness carried out on cotton linters immersed in solutions of sodium, potassium, and lithium hydroxides have shown swelling maxima for concentrations corresponding to the presence of a well-defined hydrate as seen in Table 3.6 [7].

Other studies carried out on cotton fabrics by precise measurements of thickness variations have given different results. Swelling increases with the concentration up to a constant state, without the appearance of a maximum, at least in the investigated concentration ranges: up to 300 g/liter for sodium hydroxide, 450 g/liter for potassium hydroxide, and 120 g/liter for lithium hydroxide. The percentages of swelling corresponding to these concentrations would be 120-140%, 120% and 80-85% for sodium, potassium, and lithium hydroxides, respectively. Swelling, however, has the tendency to increase again for the last two alkalis, whereas it becomes constant for a concentration of sodium hydroxide equal to 200 g/liter [28].

Normally, ions with smaller atomic volume in aqueous solutions are able to carry more water molecules than ions of larger atomic volume. The degree of swelling of cotton in these alkali hydroxides should therefore decrease in the order of decreasing hydrating power, i.e., in the order Li, Na, K. That, however, is in contradiction to the results shown in Table 3.6 [29].

Table 3.5 Maximum Swelling as a Function of Sodium Hydroxide Concentration

Cellulosic fibers	NaOH concentration (percent by weight)	Degree of swelling (%)	Swelling measurement method	References
Cotton fibers	12	40	Increase in thickness	4
	13.5	200	Increase in thickness	2
	18	78	Increase in thickness	29
Cotton linters	8-9	300	Increase in thickness	7
Cotton cloth	16-24	120-140	Increase in thickness	28
Ramie	9	1000	Increase in thickness	7
	12	170-210	Increase in thickness	4
Cuprammonium rayon	12	850	Increase in weight	30
Viscose	12	520	Increase in weight	30
	10	230	Increase in thickness	4
Polynosic	10	335	Increase in thickness	4

Table 3.6 Maximum Swelling of Cotton Linters as a Function of Alkali (*Source*: Ref. 7)

Alkali hydroxide	Concentration (percent by weight)	Degree of swelling (%)	Corresponding hydrates
NaOH	8-9	300	NaOH. $20H_2O$
KOH	45-47	480	KOH. $4H_2O$
LiOH	5-6	240	LiOH. 17 or $20H_2O$

Influence of Temperature. D'Ans and Jagër were the first to show that a decrease of temperature leads to increased swelling of cellulose fibers [32]. This increase in swelling is in accordance with Le Chatelier and Braunsch's theory, according to which the formation of alkali cellulose compounds is an exothermic process.

The variations in the swelling of cotton linters are represented in Fig. 3.2 and are computed by the measurement of the variation in

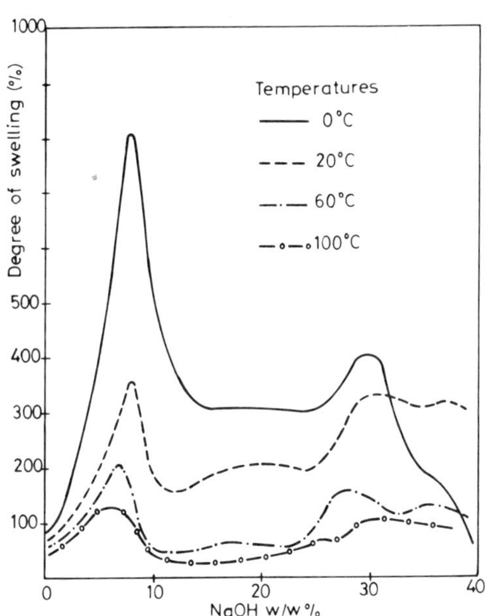

Figure 3.2 Dependence of swelling on temperature. (*Source*: Ref. 7)

thickness as a function of temperature and concentration of sodium hydroxide [7]. At 20°C, the swelling of cotton reaches a maximum on the order of 300% at a concentration of sodium hydroxide equal to 8-9% by weight. Swelling decreases next to a minimum (12-13% NaOH) and increases again up to a concentration of 25% sodium hydroxide by weight. If the temperature is decreased to 0°C, we can observe an amplification of the swelling maximum, which goes from 300 to 800% for the same concentration of sodium hydroxide (8-9%). With increasing alkali concentrations, the swelling decreases next and increases again to a maximum at a concentration of 30-32% sodium hydroxide. For higher concentrations, we observe at this temperature the crystallization of alkali hydroxide hydrates and therefore a rapid decrease in swelling. For temperatures higher than 20°C (60°C or 100°C), we can observe a decrease in swelling, the maximum being slightly shifted toward lower concentrations (7-8% NaOH by weight). At the same temperatures and for alkali hydroxide concentrations higher than 40% by weight, the percentage of swelling increases to the saturation point reaching a value on the order of 300% [7].

A similar study of the swelling of cotton fiber gave results which were relatively concordant [33]. Swelling was computed by measuring the increase of fiber width by means of a projection microscope, the degree of swelling being calculated in relation to the width of fiber in water under the same conditions. Certain differences with the preceding results are, nonetheless, to be noted.

At 0°C we can notice the existence of two maxima for concentrations on the order of 11% and 26% of sodium hydroxide by weight. At 25°C, maximum swelling takes place at a concentration of 12-13%, then swelling decreases without further increase at higher concentrations. For a temperature of 100°C, maximum swelling takes place at a 14% sodium hydroxide concentration by weight.

Measurement of Swelling. Whereas the swelling of fibers was measured for a long time exclusively by evaluating the variation of the width or thickness of the cellulosic materials, for example by microscopic means, other methods, such as the measurement of the untwisting of fibers in a swelling medium, have also been used [34-36].

The recent works dealing with swelling measurements always refer to the thickness measurements, although by means of more sophisticated and precise methods than before. We can mention, for example, the measurement of thickness variations of cotton fibers with the aid of a mobile piston and an optical system [37] or the evaluation of the thickness variations of fabrics under the action of swelling agent, likewise with the help of a mobile piston but by means of an electromagnetic system [28,38-40].

Solubility

In addition to the important swelling of cellulose fibers in alkali hydroxide solutions, we witness a certain cellulose solubility which depends on the alkali concentration and the temperature of the treatment.

Table 3.7 Maximum Solubility as a Function of Sodium Hydroxide Concentration and Temperature

Nature of fiber	Temperature (°C)	NaOH concentration (% w/w)	Solubility (% dissolved)	References
Cotton	19	16	1	42
		12-13	2	43
Viscose	-5	10	100	43
	0	10	94	44
	10	10	19	42
		10	20	43
	15	9	28	44
	18	10	16	45
	19	10	12	42
		10	14	43
	20	9	16	44
	25	10	11	44
	40	10	9	42
		10	10	43
	80	20	7	42
		20	8	43
		10	14	45

When the alkali hydroxide concentration increases, the solubility increases to a maximum and then decreases. The solubility, like swelling, is more pronounced at lower temperatures. The regenerated cellulose fibers, and in particular viscose fibers, show a much higher alkali solubility than cotton fibers. Under the effect of a sodium hydroxide solution, the low molecular weight macromolecules are extracted from viscose fibers so that the degree of polymerization of residual fibers increases [41].

The maximum solubilities of cotton and viscose fibers are summarized in Table 3.7 as a function of sodium hydroxide concentration and temperature.

Maximum solubility of viscose fibers occurs at room temperature for sodium hydroxide concentrations on the order of 9-10% by weight (100-110 g/liter). At this concentration the hydrated ion pairs $NaOH \cdot 20H_2O$ are in greater proportion. These ions contribute to cotton fiber swelling but only enter the amorphous regions without inducing any increase in the solubility of the cotton fibers. On the other hand, in the case of viscose fibers these ions penetrate the amorphous regions as well as the higher lateral order regions, causing an unlimited swelling which leads to the solubilization of macromolecules [46]. Another explanation for the behavioral differences between cotton and viscose fibers is based on the shorter length of the macromolecular chains of the viscose fiber and on the higher proportion of carboxylic groups [47].

At sodium hydroxide concentrations higher than 13% by weight (150 g/liter), the solubility of viscose fibers is considerably less. In fact, the solvated dipole hydrates and even the hydrated ion pairs $NaOH \cdot 12H_2O$ present at these concentrations have a lower swelling ability, which reduces the risks of the solubilization of viscose. The dissolution of viscose is very important in sodium hydroxide solutions, but it is less so in potassium hydroxide solutions, as seen in Fig. 3.3.

From a practical point of view, the process of dissolving viscose fibers at a critical (approximately 100 g/liter) sodium hydroxide

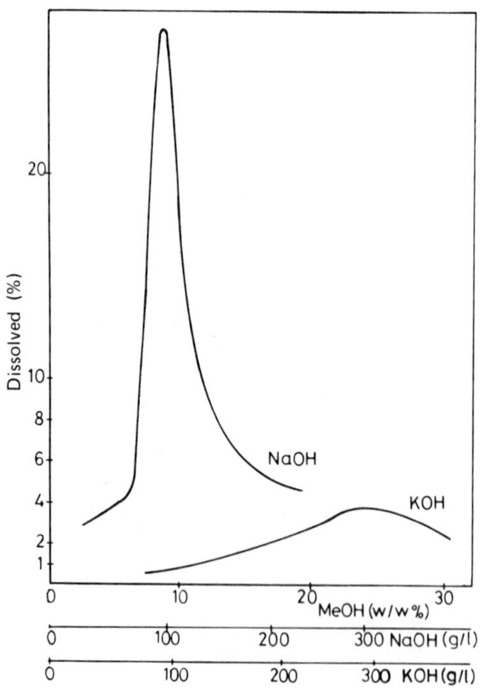

Figure 3.3 Solubility of viscose in alkaline solutions at 15°C. (*Source*: Ref. 44.)

concentration creates problems when mercerizing blended textile materials containing viscose fiber (for example, cotton-viscose blends). The impregnation step of mercerization is carried out at a concentration of about 300 g/liter (24% by weight), where the viscose solubility is low. However, during subsequent rinsing the sodium hydroxide concentration passes through the critical value of 100 g/liter, leading to a significant solubilization of fibers and a considerable decrease in the mechanical properties of the textile materials. To avoid this mechanical degradation, great care must be taken in the rinsing process. It should be done at high temperatures, which lower the solubility of viscose, in the presence of electrolytes such as sodium chloride, the addition of which decreases the swelling power of sodium hydroxide hydrates by affecting their degree of hydration [45,48,49].

Concerning the other types of regenerated cellulose fibers, their solubility in alkali solutions is also higher than that of cotton fibers. The solubility is higher for fibers with high wet modulus than for polynosic fibers [50-55]. After treating high-wet-modulus fibers in a slack state in sodium hydroxide solutions we can observe a considerable loss of tensile strength, on the order of 40% for a concentration of 5% by weight (50 g/liter NaOH) and of 60% for a concentration of 23.5% by weight (300 g/liter NaOH) [53]. However, if the treatment is conducted under tension sufficient to avoid shrinkage, the mechanical degradation resulting from the solubility is less pronounced [51,53].

The high resistance of polynosic fibers to sodium hydroxide solutions is a fundamental property which distinguishes them from other regenerated cellulose fibers and which results from the higher value of their degree of polymerization.

When polynosic fibers are treated in sodium hydroxide solutions of 5 and 23.5% by weight, the losses in tensile strength are, respectively, 15 and 30%, i.e., less than for high-wet-modulus fibers. The losses are also reduced if the treatment is conducted under increasing tension [53].

2.4 Degradation of Cellulose in Alkali Solutions

The degradation of cellulose in alkali solutions varies depending on the presence or absence of oxygen. In the absence of air, cellulose is slowly attacked by hot alkali in a stepwise fashion at the reducing end of the cellulosic chain. In this case we speak of hot alkali solubility [56]. In the presence of air, the degradation can be very serious: an oxidation phenomenon adds to the action of alkali ions, inducing a profound depolymerization of the cellulose chain [57]. This reaction is called alkali cellulose autooxidation.

Hot Alkali Solubility

During hot alkali treatment of cellulose containing reducing end groups, the cellulose molecules are subjected to endwise degradation which results in a loss of material. This degradation explains the fact that the weight loss of cellulose material during scouring is greater than the total quantity of extractable impurities. This loss of weight increases with the temperature of the alkali treatment [58]. A reaction scheme of this kind of degradation can be illustrated in the following way [59].

Formation of peeling-off centers on the cellulose, i.e., aldehyde groups.

Progressive stepwise peeling-off reaction of anhydroglucose units by β-alkoxy elimination reaction from the reducing ends of the chains in the amorphous regions of the fibers. The peeling-off mechanism involves not only the lower order regions associated with the crystalline regions, but those associated laterally with their fringes and also those limiting the macrofibrils [60].

Formation of carbonyl groups containing colored products proportional to the active aldehydes present in the cellulose and to the number of glucose units peeled off (yellowing) [61].

Stopping reaction: formation of metasaccharinic acid units at the end of the chains [62-64]. Thus the degradation of cellulose does not go to completion because the metasaccharinic end groups are stable to alkali.

Alkali Cellulose Autooxidation

When cellulose is converted to alkali cellulose and exposed to air, oxidative degradation occurs with chain scission and a decrease in the degree of polymerization. This reaction is important in the textile field, e.g., in the aging of soda cellulose or treatment of pulp with hot alkali, which finds application in the manufacture of viscose. This way of lowering the molecular weight in the viscose process was first utilized in the late 1890s [65]. A number of researchers have already studied this reaction, either to define its industrial conditions or to try to clarify its mechanism [57,66-74]. It was confirmed that hydrogen peroxide is always present during cellulose autooxidation. Entwistle, Cole, and Wooding [66] advanced a theory based on free radical intermediates, as seen in Fig. 3.4.

The attack by oxygen on some hydrogen atoms of the cellulose (compound I) produces a radical II which takes up oxygen to form a peroxide radical III. This peroxide radical reacts with cellulose itself to give hydroperoxide IV and radical II (in the propagation step).

Hydroperoxide IV gives a hydroxyl radical and a new radical V, which may abstract hydrogen from cellulose to give compound VI and the original radical II. The hydroxyl radical may attack cellulose to give new type II radicals (in autocatalysis). Hydroxyl radicals may also combine to form hydrogen peroxide. Compound VI finally loses water with the formation of a carbonyl group in the cellulose chain (compound VII). If the carbonyl groups introduced are located in a

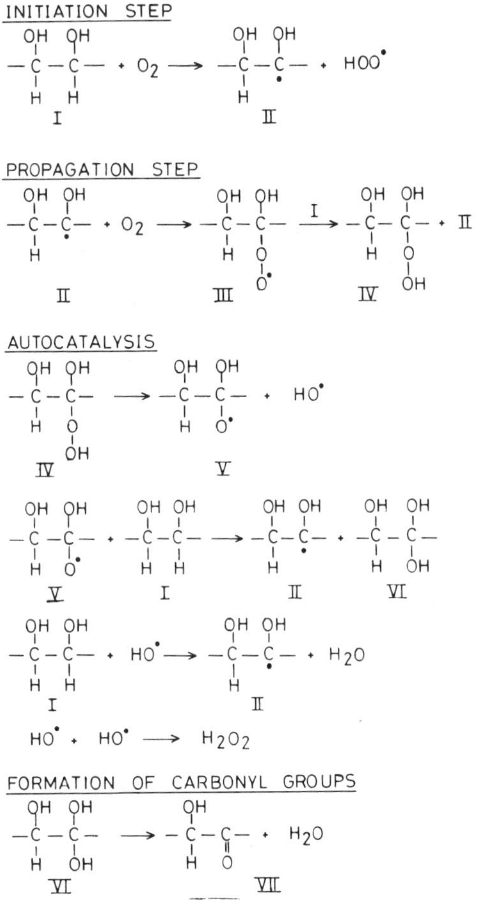

INITIATION STEP

```
 OH OH               OH OH
 |  |                |  |
-C--C-  + O2 ---->  -C--C-  + HOO•
 |  |                |  •
 H  H                H
  I                    II
```

PROPAGATION STEP

```
 OH OH             OH OH             OH OH
 |  |              |  |       I      |  |
-C--C-  + O2 ---> -C--C-   ------>  -C--C-  + II
 |  •              |  |              |  |
 H                 H  O              H  O
                      |                 |
                      O•                OH
  II               III               IV
```

AUTOCATALYSIS

```
 OH OH              OH OH
 |  |               |  |
-C--C-    ------>  -C--C-   + HO•
 |  |               |  |
 H  O               H  O•
    |
    OH
 IV                   V

 OH OH        OH OH           OH OH       OH OH
 |  |         |  |            |  |        |  |
-C--C-   +   -C--C-   --->   -C--C-   +  -C--C-
 |  |         |  |            |  •        |  |
 H  O•        H  H            H           H  OH
  V             I               II          VI

 OH OH                  OH OH
 |  |                   |  |
-C--C-  + HO• ---->    -C--C-   + H2O
 |  |                   |  •
 H  H                   H
  I                       II

HO• + HO• ---> H2O2
```

FORMATION OF CARBONYL GROUPS

```
 OH OH              OH
 |  |               |
-C--C-   ------>   -C--C-   + H2O
 |  |               |  ||
 H  OH              H  O
  VI                   VII
```

Figure 3.4 Free radical mechanism of alkali cellulose autooxidation.

certain position in the cellulosic chain, scission occurs by β-alkoxy elimination, as seen in Fig. 3.5 [75].

However, the presence of free radical intermediates has never been proven; Mattor [76,77] proposed an ionic mechanism for the autooxidation of alkali cellulose, as seen in Fig. 3.6. Reducing groups of cellulose (aldehyde groups), when placed in a strong alkali solution, form Cannizzaro intermediates such as IX and X which are autooxidizable. The initiation step of the autooxidation reaction consists of the formation of the perhydroxyl ion. The propagation step is the attack of cellulose by the perhydroxyl ion with rupture of the cellulose chain, following a complex mechanism which is still unknown and is similar to the degradation mechanism of cellulose during bleaching with hydrogen peroxide. However, the two following modes of attack may be cited:

Nucleophilic substitution at the glucosidic linkage by the perhydroxyl ion
Oxidation of any hydroxyl to a carbonyl group and subsequent β-alkoxy elimination

The attack on cellulose by the perhydroxyl ion could also generate new aldehyde groups which contribute to the formation of new perhydroxyl ions. This could explain the autocatalysis of alkali cellulose oxidation. Alkali cellulose autooxidation proceeds only under strong alkaline conditions, i.e., at sodium hydroxide concentrations above 40 g/liter [57]. When the cellulose is saturated in a concentrated solution of sodium hydroxide then squeezed and put in contact with air, the number of cellulosic chain scissions increases rapidly with an increase in sodium hydroxide concentration up to 250 g/liter and then in a less noticeable way, whereas the degree of polymerization decreases accordingly [78]. Although the above autooxidation reactions occur at high sodium hydroxide concentrations, it is probable that the attack of oxygen on cotton saturated with alkali liquor at concentrations used in scouring

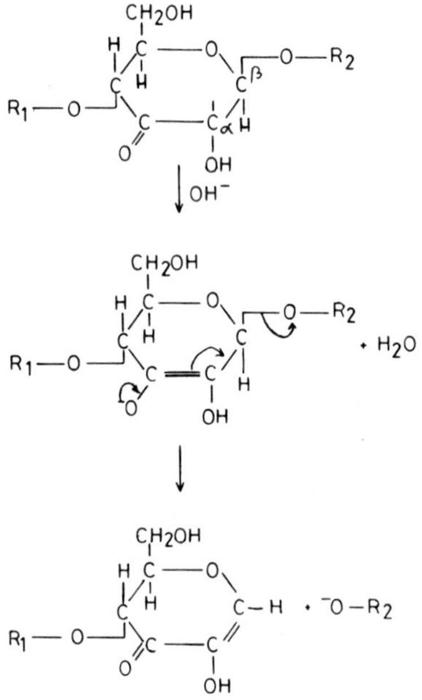

Figure 3.5 β-alkoxy elimination.

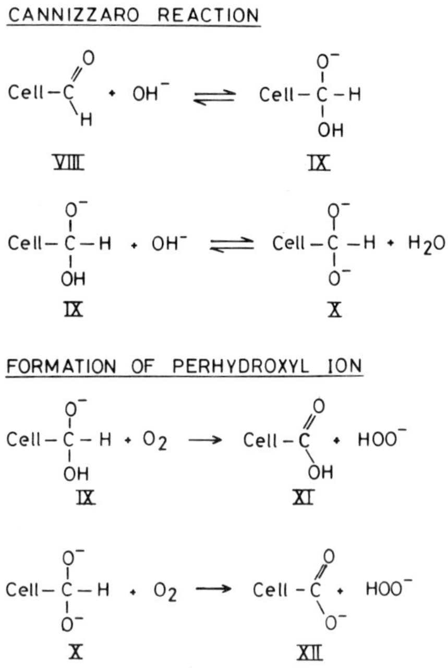

CANNIZZARO REACTION

$$\underset{\text{VIII}}{\text{Cell}-\overset{O}{\overset{/\!/}{C}}\diagdown H} + OH^- \rightleftharpoons \underset{\text{IX}}{\text{Cell}-\overset{O^-}{\overset{|}{\underset{OH}{\underset{|}{C}}}}-H}$$

$$\underset{\text{IX}}{\text{Cell}-\overset{O^-}{\overset{|}{\underset{OH}{\underset{|}{C}}}}-H} + OH^- \rightleftharpoons \underset{\text{X}}{\text{Cell}-\overset{O^-}{\overset{|}{\underset{O^-}{\underset{|}{C}}}}-H} + H_2O$$

FORMATION OF PERHYDROXYL ION

$$\underset{\text{IX}}{\text{Cell}-\overset{O^-}{\overset{|}{\underset{OH}{\underset{|}{C}}}}-H} + O_2 \longrightarrow \underset{\text{XI}}{\text{Cell}-\overset{O}{\overset{/\!/}{C}}\diagdown OH} + HOO^-$$

$$\underset{\text{X}}{\text{Cell}-\overset{O^-}{\overset{|}{\underset{O^-}{\underset{|}{C}}}}-H} + O_2 \longrightarrow \underset{\text{XII}}{\text{Cell}-\overset{O}{\overset{/\!/}{C}}\diagdown O^-} + HOO^-$$

Figure 3.6 Ionic mechanism of alkali cellulose autooxidation. (*Source*: Ref. 77.)

(kier boiling) follows a mechanism similar to that suggested for alkali cellulose.

The temperature has a considerable influence on the speed of autooxidation of alkali cellulose. The attack of oxygen at 20°C is slow. It becomes more rapid starting at 40-50°C [57,58], and the speed of autooxidation is 1000 times higher at 95°C than at 20°C [66].

Finally, regarding the influence of the type of alkali hydroxide on cellulose autoxidation, potassium hydroxide is much more active than sodium hydroxide and lithium hydroxide [79], a fact that is in agreement with the values of the degree of swelling shown in Table 3.6; in fact, the intramicellar swelling plays an important role in autooxidation processes [78].

3. SCOURING OF COTTON

3.1 Introduction

It may be said that good scouring is the foundation of successful finishing. In fact, scouring is a purifying treatment of cotton α-cellulose, which contains impurities such as waxes, pectins, hemi-celluloses, and mineral salts. Native cotton contains a certain proportion of natural waxes which bestow a soft touch on the fiber but are undesirable for the subsequent dyeing and finishing operations. These products are consequently retained as long as they are advantageous (e.g., for lubricating action in spinning) and then must be eliminated to ensure water absorbency. The water absorbency is necessary for the uniformity of bleaching, dyeing, printing, and chemical finishing treatments.

It should be noted that there is no correlation between the water absorbency of cotton fiber and its natural wax content. In fact, it is mainly the distribution of the residual wax that determines water absorbency. Thus, a cotton fabric subjected to solvent extraction has a very low water absorbency compared to the same fabric simply immersed in boiling water and dried at room temperature. In the latter case, the wax melts and forms droplets. After drying in the air, the wax coating is not continuous and the fiber becomes absorbent. On the other hand, if we dry the same fiber at a temperature higher than the melting point (60-70°C) of the wax, droplets of the wax melt and reform a continuous film around the fiber, rendering it hydrophobic [80]. In practice, it is not possible to dry cotton goods at room temperature; it is necessary, therefore, to reduce the wax content to a level at which it can no longer have a continuous protective film formation around the fiber during drying.

Mineral salts constitute another impurity to which attention should be paid. Mineral salts such as alkali earth metal salts may be neglected in exhaustion dyeing; however, in continuous dyeing with a bath ratio less than 1/1, the presence of alkali earth metal salts changes soft water to hard water, rendering certain dyes insoluble which will remain attached to the fiber surface [81]. For pad dyeing processes it is therefore indispensable to have a pretreatment capable of eliminating alkali earth metal salts from the fiber. Besides the undesirable waxes and mineral salts, the cotton goods also contain a certain proportion of lignified substances, consisting of the residues of seeds, husks, and leaves which have not been completely removed mechanically during yarn fabrication. These heavily colored substances must also be eliminated during the pretreatment.

In fact, scouring is the only industrial method capable of removing cotton impurities almost totally. The scouring process consists of an alkali treatment in the presence of wetting and sequestering agents in order to convert the impurities, other than natural coloring matter, into products which can be removed by aqueous washing.

3.2 Impurities of Cellulosic Materials

The overall composition of cotton fibers is indicated in Table 3.8. The percentages of impurities that are found depend on the origin of the cotton (i.e., kind of soil, climate, and plant variety). The amount of

Table 3.8 Composition of Mature Cotton Fibers

Constituents	Percent of dry weight	References
α-cellulose	88.0-96.5	82,83,84
Protein	1.0-1.9	82,84,85,86
Wax	0.4-1.2	82,83,84,85,86,87,88
Ash (inorganic salts)	0.7-1.6	82,84,86
Pectin	0.4-1.2	82,86,87
Others (resins, pigments, hemicellulose, sugars, organic acids, incrusted ligneous substances)	0.5-8.0	

cotton fiber impurities also varies according to the degree of fiber maturity. Immature fibers include a higher proportion of impurities, as seen in Table 3.9. Cotton impurities are located largely on the outer side of fibers (see Table 3.10); their quantity is higher when the fibers are finer, that is to say when their specific surface area is larger [89,90].

Table 3.9 Influence of Maturity on the Impurities of Cotton Fibers (in Percent of Dry Weight)

	U.S. cotton[a]		USSR cotton[b]	
Constituents	Mature	Immature	Mature	Immature
Wax	0.45	1.14	0.65	1.65
Protein	1.01	2.02	1.58	3.36
Ash	0.79	1.32	1.17	1.56
Pectin	0.58	1.26	0.68	1.43

[a]Ref. 84.
[b]Ref. 86.

Table 3.10 Impurities in Cotton Fibers (in Percent of Dry Weight) by Location

Constituents	Total fiber	Primary wall
α-cellulose	94	54
Protein	1.3	14
Pectin	1.2	9
Wax	0.6	8
Ash	1.2	3
Others	1.7	12

Source: Ref. 90.

Cotton Waxes

Waxes with their hydrophobic character, have an essential protective role in plants. Waxes constitute the impurity which is most difficult to eliminate from cotton fibers. Their quantity varies between 0.4 and 1.2% of the weight of anhydrous mature cotton and depends on the fiber origin. The measured quantity of waxes also varies depending on the solvent used for their extraction (ethanol, ether, benzene, methylene chloride, chloroform, carbon tetrachloride, etc.) as well as on the length of extraction [85,91].

Cotton waxes appear in the form of a very complex substance whose principal constituents are high molecular weight alcohols and fatty acids in free or esterified form. The melting point of cotton waxes can vary between 68 and 80°C [86,92,93]. Their constituents

Table 3.11 Composition of Cotton Wax

Main constituents	Percent of dry weight
Unsaponifiable matter	50-77
Alcohols	40-52
Acids	23-47
Hydrocarbons	7-13
Sterols and polyterpenes	3-18

Sources: Refs. 85,86,92,94.

Table 3.12 Percentage of Unsaponifiable Matter

Origin of cotton	Unsaponifiable matter (percent of dry weight)	References
North America	52	86
	58-77	92
	66-68	95
	58	93
	62-68	96
	60-62	97
South America	53-64	92
	65-68	97
USSR	62	86
	52	97
Egypt	58	86
	50-58	92
	57-68	94
	57	96
	48-54	97
Turkey	61	97

are referred to in Table 3.11. The percentage of unsaponifiable matters (fatty alcohols, hydrocarbons, sterols, polyterpenes) can vary depending not only on the cotton origin and harvest but also on the samples studied as seen in Table 3.12. One elementary analysis of cotton waxes gave the following proportions [85].

Carbon	80.38%
Hydrogen	14.51%
Oxygen	5.11%

The main fatty acids of cotton waxes, in the free or combined form, are listed in Table 3.13.

Palmitic and stearic acids are the fatty acids present in the highest proportion. All fatty acids contain an even number of carbon atoms. Certain names assigned to fatty acids isolated from cotton waxes have been given to products that are not chemical entities but mixtures [98], such as cerotic, montanic, melissic, gossypic, and carnaubic acids.

Table 3.13 Fatty Acids Found in Cotton Wax

Chemical formula	Common name	Systematic name
$C_{15}H_{31}COOH$	Palmitic	Hexadecanoic
$C_{17}H_{33}COOH$	Oleic (unsaturated)	Octadec-9-enoic
$C_{17}H_{35}COOH$	Stearic	Octadecanoic
$C_{19}H_{37}COOH$	Gadoleic (unsaturated)	Eicos-9-enoic
$C_{19}H_{39}COOH$	Arachidic	Eicosanoic
$C_{21}H_{43}COOH$	Behenic	Docosanoic
$C_{23}H_{47}COOH$	Lignoceric (Carnaubic)	Tetracosanoic
$C_{25}H_{51}COOH$	(Cerotic)	Hexacosanoic
$C_{27}H_{55}COOH$	(Montanic)	Octacosanoic
$C_{29}H_{59}COOH$	(Melissic) (Gossypic)	Triacontanoic
$C_{31}H_{63}COOH$	Laccer acid	Dotriacontanoic
$C_{33}H_{67}COOH$	Ghedda acid	Tetratriacontanoic

Source: Refs. 86,93,94.

High molecular weight alcohols found in cotton waxes are listed in Table 3.14. The C_{24}-C_{30} alcohols are present in the largest quantity and are mainly gosspyl and montanyl alcohols, which are not pure alcohols but are mixtures of two or more alcohols [98]. Among the alcohols present in cotton glycerol is found as a low-molecular-weight alcohol and is exclusively combined to fatty acids in the form of esters (glycerides).

Besides alcohols and fatty acids, which are the principal constituents of cotton waxes, we can count a certain number of other products, such as phytosterols and polyterpenes (sitosterol $C_{29}H_{50}O$, sitosterol glucoside or sitosterolin $C_{35}H_{60}O_6$, lupeol $C_{30}H_{50}O$, α and β-amyrin $C_{30}H_{50}O$), hydrocarbons (heptacosane $C_{27}H_{56}$, triacontane $C_{30}H_{62}$, untriacontane $C_{31}H_{64}$, dotriacontane $C_{32}H_{66}$) and colored resinous products [85,86,93,94,99].

Cotton waxes are located mainly in the primary wall but are also present in the secondary wall [100,101]. Two possibilities have been postulated concerning the linkage between cellulose (or pectins) and waxes (see Fig. 3.7). According to Hess [102], a phosphatide, in particular lecithin, plays the connecting role between the hydrophilic cellulose or pectins and the hydrophobic wax. Von Hornuff and Richter [86] offer another kind of connection based on substances combining amino acids, fatty acids, and eventually glucosidic units.

Proteins

Cotton proteins consist of protoplasmic residues, but little is known about their detailed composition. The coloration of cotton could be connected to its protein content [85,103,104], which varies between 1.0 and 1.9% and which is estimated by multiplication of the nitrogen content by 6.25-6.30. The proteins are mainly concentrated in the primary wall of the fibers (see Table 3.10) but they appear also in the lumen [105]. The brown color which appears during cotton scouring could be due in part to a reaction between proteins and carbohydrates in the alkaline medium [106,107].

Pectic Substances

The main constituent of pectic substances in cotton is a poly-D-galacturonic acid in which about 85% of the carboxyl groups are methylated. Pectin content in cotton fibers is on the order of 0.4-1.2%. These substances are present in the form of insoluble salts of calcium, magnesium, and iron [108,109]. These bivalent metal salts (pectates) are solubilized in sodium hydroxide solutions, either by the formation of

Table 3.14 Fatty Alcohols Found in Cotton Waxes

Chemical formula	Common name	Systematic name
$C_{24}H_{49}OH$	Lignoceryl alcohol (carnaubyl alcohol)	Tetracosanol
$C_{26}H_{53}OH$	Ceryl alcohol	Hexacosanol
$C_{28}H_{57}OH$	(Montanyl alcohol)	Octacosanol
$C_{30}H_{61}OH$	(Gossypyl alcohol)	Triacontanol
$C_{32}H_{65}OH$		Dotriacontanol
$C_{34}H_{69}OH$		Tetratriacontanol
$C_{28}H_{56}(OH)_2$		Octacosandiol
$C_{30}H_{60}(OH)_2$	Coceryl alcohol	Triacontandiol
$C_3H_5(OH)_3$	Glycerol	Propanetriol

Sources: Refs. 86, 93, 94.

metapectic acid or by an exchange between bivalent metallic ions and sodium cations [86].

Inorganic Matter (Ash)

The quantity of ash is on the order of 0.7-1.6% by weight of anhydrous cotton. This quantity may be decreased by 85% by simply boiling the cotton in water [108]. The composition of cotton ash is given in Table 3.15.

Among the inorganic substances it is appropriate to mention the phosphorus content. Expressed in the form of P_2O_5, it can vary between 0.05% and 0.25% depending on the origin of the fibers, this content being higher for Egyptian and Russian cottons [86,110]. Phosphorus is present in the form of organic and inorganic compounds which are for the most part soluble in hot water but which can become

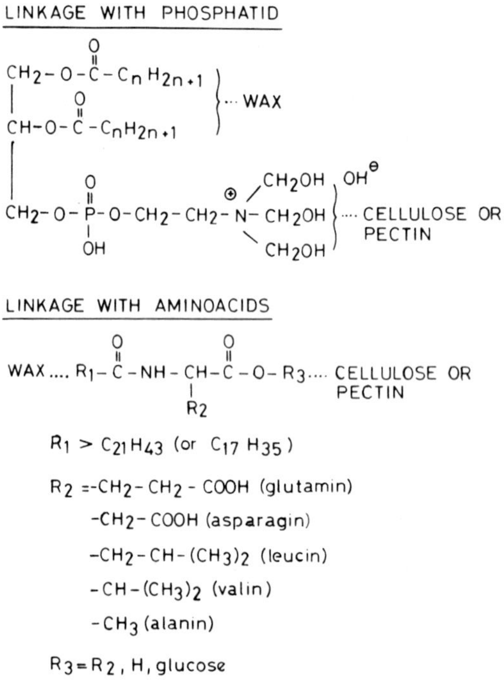

LINKAGE WITH PHOSPHATID

$CH_2-O-C(=O)-C_nH_{2n+1}$ } ... WAX

$CH-O-C(=O)-C_nH_{2n+1}$

$CH_2-O-P(=O)(OH)-O-CH_2-CH_2-N^{\oplus}(CH_2OH)_3$, $OH^{\ominus}$ } .... CELLULOSE OR PECTIN

LINKAGE WITH AMINOACIDS

WAX .... $R_1-C(=O)-NH-CH(R_2)-C(=O)-O-R_3$ .... CELLULOSE OR PECTIN

$R_1 > C_{21}H_{43}$ (or $C_{17}H_{35}$)

$R_2 = -CH_2-CH_2-COOH$ (glutamin)

$-CH_2-COOH$ (asparagin)

$-CH_2-CH-(CH_3)_2$ (leucin)

$-CH-(CH_3)_2$ (valin)

$-CH_3$ (alanin)

$R_3 = R_2$, H, glucose

Figure 3.7 Linkage between wax and cellulose or pectin. (*Source*: Ref. 86, 102.)

insoluble in the presence of alkali earth metals. Therefore, the use of hard water during scouring can precipitate alkali earth metal phosphates on the fiber instead of eliminating them [86].

Table 3.15 Composition of Cotton Ash (Percent by Weight)

Constituents	Ref. 95	Ref. 84	Ref. 108	
Na_2O		0.25-0.66	7	
NaCl				7.17
Na_2CO_3	3.4			
K_2O		45.6-48.2	34	
KCl	10.2			
K_2CO_3	33.2			34.80
K_2SO_4	13.0			20.46
K_2SiO_3				3.44
$K_2P_2O_7$				8.32
CaO		7.2-7.6	11	
$CaCO_3$	20.3			6.69
$CaSO_4$				5.16
$CaSiO_3$				2.77
MgO		9-10.9	6	6.60
$MgCO_3$	7.8			
$Mg_3(PO_4)_2$	8.7			
Fe_2O_3	3.4	0.20-0.36	2	4.59
Al_2O_3		0.31-0.73	2	
SiO_2		0.76-0.75	5	
P_2O_5		4.2-4.4	5	
SO_3		7.04-8.24	4	
Cl		1.2-2.1	4	
CO_2		20.3-19.8	20	

Other Substances

Besides waxes, proteins, pectins, and mineral salts, we also find a certain number of other compounds present in larger or smaller quantities; thus hemicelluloses, which are cross-linked polysaccharides, are more soluble in alkalis and more easily hydrolyzable in acids than cellulose. Cotton also contains free reducing sugars (hexoses and pentoses, and, specifically, free glucose) which constitute the base units of hemicellulose. Finally, like other cotton constituents, organic acids (citric, malic acid), encrusted ligneous substances, and colored pigments are also to be found.

3.3 Physicochemical Aspects of the Scouring Process

Various processes can be used to carry out scouring successfully. These processes can be distinguished by whether the operation takes place in a noncontinuous or in a continuous way. The noncontinuous scouring process consists essentially of circulating hot alkali liquor through the cloth which is formed into a rope and piled in a vessel known as a kier (from which the name kiering or kier boiling stems). In the continuous process higher productivity is based on the impregnation of the fabric in an alkali solution followed by a steam aging stage, the fabric being either in rope form, or more recently in open-width form.

The control of the scouring efficiency can take place in different ways:

- Measurement of weight loss
- Measurement of protein content (nitrogen analysis)
- Measurement of residual wax content
- Measurement of methylene blue absorption (removal of pectic substances)
- Measurement of copper number (removal of hemicelluloses and sugars)
- Practical test of absorbency (The practical test consists of measuring the time of absorption of a drop of water allowed to remain in contact with the scoured cloth [111]. If the drop is absorbed in less than 3 sec, the cloth is judged suitable for bleaching and dyeing. In the case of printing, the time of absorption of the drop of water must be lower than 1 sec.)

Effects of Alkaline Treatment

When cotton fabric is submitted to a scouring treatment in a warm sodium hydroxide solution, we observe a certain number of reactions that render the cotton impurities soluble:

- Proteins are hydrolyzed with the formation of soluble sodium salts of amino acids.

Pectins are changed into water-soluble salts of pectic or metapetic acid.

Hemicelluloses as well as cellulose fractions with a low polymerization degree are dissolved.

Inorganic substances are partially dissolved.

The saponifiable parts of waxes (esters, glycerides, fatty acids) are converted into soaps. The unsaponifiable parts of waxes (alcohols, hydrocarbons), which are molten at the scouring temperature are emulsified by the soaps formed from the saponifiable parts.

This emulsification process plays a considerable role in the removal of cotton waxes during scouring. It is not necessary to eliminate the waxes quantitatively, but the residual wax quantity must be insufficient to re-form a continuous layer over the fiber surface [112].

Although sodium hydroxide is used generally for scouring, sodium carbonate and calcium hydroxide should also be mentioned as scouring agents. For example, the lime soda sequence is a process which was used in the past and consists of boiling the cloth with milk of lime. This treatment induces the formation of insoluble lime soaps. The second step or subsequent boil in sodium carbonate allows the lime soap to be transformed into calcium carbonate and water-soluble alkali soaps. In certain cases a mixture of sodium carbonate and sodium hydroxide or a mixture of sodium carbonate and soap (the latter system is milder than the sodium hydroxide) is used for scouring delicate clothes such as colored woven goods.

The scouring process, based on the reaction between cotton impurities and alkali hydroxide, implies a certain alkali consumption which determines the minimum concentration of sodium hydroxide to be used. When sodium hydroxide is brought into contact with the cotton cloth, some of the alkali is absorbed, since the hydroxyl groups of cellulose have a weak acidic character [113]. So, at pH 13.3 cellulose absorbs about 1% or 10 g/kg of sodium hydroxide. Alkali is also required to neutralize the carboxyl groups of the pectins (about 0.5%, or 5 g per kg of cotton). Neutralization of amino acids obtained by hydrolysis of proteins uses around 1% sodium hydroxide, or 10 g per kg of cotton. If we consider the quantity of sodium hydroxide necessary for the saponification of waxes and the importance of retaining an excess of sodium hydroxide in order to maintain sufficient alkalinity, it is evident that the minimum concentration of sodium hydroxide should be on the order of 3% by weight of fabric to be scoured. This minimum concentration of 3% has also been obtained from measurements of the water absorbency of scoured goods as a function of sodium hydroxide concentration [114].

In noncontinuous processing (kiering), the liquor/fabric ratio is equal to 3/1; consequently, the use of a 10-20 g/liter sodium hydroxide solution means a quantity of about 3-6% of the weight of fabric. In these conditions the length of treatment ranges from 2 to 12 hr at temperatures between 100 and 130°C. In more recent continuous scouring processes based on a 100% pickup, the sodium hydroxide concentration should be a minimum of 30 g/liter. However, the steaming time is necessarily less than in the batch process to attain higher productivity.

After scouring some of the cotton impurities are dissolved while the rest are dispersed in the solution. It is therefore important not to lower the temperature of the bath abruptly in order to avoid the breakdown of the emulsion and the precipitation of the impurities onto the cotton. It is necessary to carry out a hot progressive rinsing while gradually decreasing the temperature. Even after an efficient rinsing, cotton fabrics retain a certain quantity of sodium hydroxide, which necessitates a neutralizing treatment in an acid solution.

Influence of Temperature and Pressure

Because only relatively diluted solutions of alkali are used to scour cotton, particularly in a noncontinuous process, the rate of saponification of waxes is relatively low but can be increased by operating at higher temperatures, i.e., under pressure. As seen in Fig. 3.8, the amount of residual wax decreases significantly as the temperature (pressure) of kier boiling increases [94]. In practice, the noncontinuous kiering process takes place in a vessel under a saturated steam pressure of

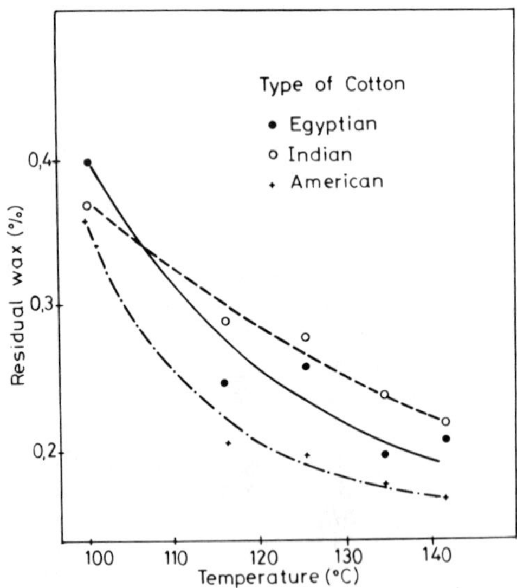

Figure 3.8 Influence of scouring temperature on the percent of residual wax. Treatment for 6 hr at NaOH concentration of 10 g/l. (*Source*: Ref. 94.)

up to 30 lb/in^2, which corresponds to a temperature of about 130°C and allows for a decrease in the time of treatment.

Increasing the temperature has the same effect in the case of the open-width scouring process, provided a steamer for continuous treatment under pressure is available. With a sodium hydroxide solution of 40-60 g/liter it is therefore possible to decrease the time of post-impregnation steaming to about 2 min if we operate at a temperature of 130-135°C [115-120].

Effect of Auxiliary Products

In the scouring process, the addition of a surface-active agent is recommended to aid the caustic soda in its functions. First, since the gray fabric is hydrophobic, the surface-active agent is added to the scouring bath in order to allow even and rapid wetting of the cloth by the alkali solution. Moreover, the surface-active agent should have detergent action, with its dispersing and emulsifying properties facilitating the suspension and removal of the impurities from the fabric. The use of surface-active agents is especially important for the continuous process, where the short time of treatment necessitates a rapid elimination of waxes and other impurities. These products must be chosen according to the following criteria [121]:

Stability in a hot alkaline medium. (However, the wetting and dispersing properties of the surface-active agents decrease when their alkali stability increases and their choice must also take into account the alkali concentration.)

Good electrolyte stability.

Wetting and dispersing actions; the surface-active agents do not always have wetting and dispersing properties simultaneously. It is consequently usual to combine two products in order to obtain more desirable result.

Good removal during rinsing and weak foaming power.

Among the most often used surface-active agents, we can cite certain anionic products such as sodium alkyl sulfates and alkyl aromatic sulfonates. Such anionic products are very often combined with nonionic products such as polyethoxylated compounds.

In the hot alkaline medium cellulose can be damaged if we operate in the presence of oxygen. In noncontinuous processes the scouring is therefore carried out in a closed vessel, and air should be carefully removed from the kier.

However, to avoid all risks of damage to the cotton, a small quantity of a reducing compound should be added to the scouring bath [81]. In the case of a continuous process, the use of a reducing compound is also advisable considering that the aging chambers cannot be tightly closed. Sodium bisulfite and sodium hydrosulfite are the most often used reducing agents.

The scouring process requires the use of soft water, because the use of hard water would cause the insolubilization of soaps formed from waxes in the presence of alkali earth ions (calcium, magnesium). It must be considered that cotton fibers contain calcium and magnesium salts (pectin salts), which are freed during the alkali treatment and can also contribute to the insolubility of soaps and at the same time remain attached to the goods in the form of hydroxides. Thus, they might disturb subsequent operations such as bleaching and dyeing. In the continuous scouring process, we have a scouring bath of hard water even when distilled water is used. In fact, the padding step of the fabric is carried out with a bath ratio of less than 1/1, which brings about a rather sizable concentration of calcium and magnesium ions extracted from the fibers [122].

It is, therefore, necessary to add a sequestering agent to the scouring bath in order to eliminate calcium and magnesium ions even when soft water is used. The sequestering agent also allows for the elimination of iron, which can originate problems during subsequent bleaching with hydrogen peroxide. The oldest sequestering agents are the inorganic polyphosphates, which in addition to their chelating effect also have dispersing properties and prevent soil redeposition on textile goods. However, these products are not stable in alkali solutions at high temperatures and can only be used in sodium hydroxide solutions of less than 15 g/liter concentration. For higher concentrations, as are often used, they are hydrolyzed to orthophosphates, and thus lose their dispersing and sequestering characteristics.

Organic sequestering agents such as aminocarboxylates [nitrilotriacetic acid (NTA), ethylenediaminotetraacetic acid (EDTA), diethylenetriaminopentaacetic acid (DTPA)] and hydroxycarboxylates (gluconic and glucoheptonic acids) are much more stable in hot alkali solutions but, unlike the polyphosphates, they do not contribute to detergency. A new kind of sequestering agent, however, allows a complex formation with calcium and magnesium ions and at the same times possesses properties helping in the removal of soil and preventing soil redeposition. These agents are hydrolytically stable in the hot alkaline medium. The chemical constitution of these sequestering agents (organophosphonates) is given in Fig. 3.9 [123].

Commercial scouring assistants are generally composed of a sequestering agent, a fiber-protective reducing compound, and an extraction agent that accelerates the removal of cotton impurities, therefore permitting a reduction in the scouring time [124-127].

For the purpose of accelerating the scouring process, wetting agents are sometimes used in conjunction with high-boiling solvents (cyclohexanol, methylcyclohexanol, tetrahydronaphthalene, decahydronaphthalene).

Properties of Scoured Cotton

The main changes in cotton goods during the scouring process are loss in weight (about 5-10%), loss in length due to shrinkage during

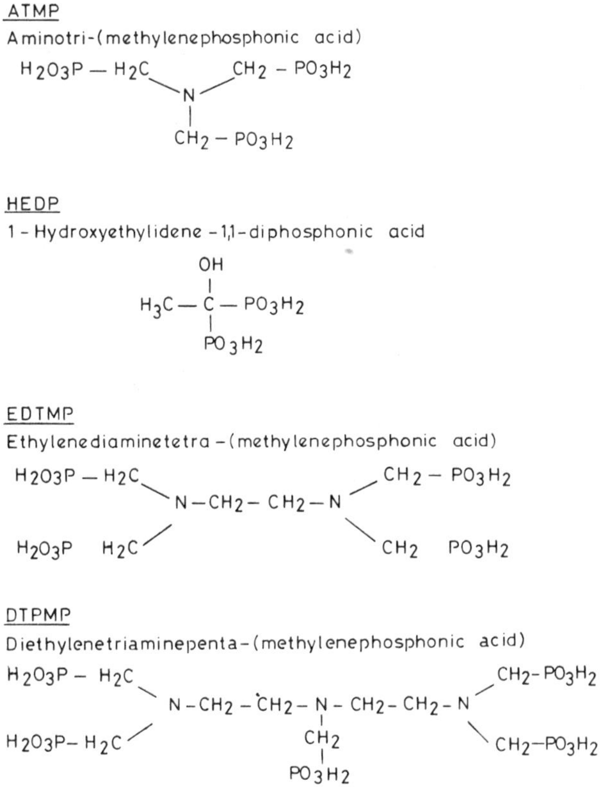

ATMP

Aminotri-(methylenephosphonic acid)

$$H_2O_3P - H_2C\diagdown N \diagup CH_2 - PO_3H_2$$
$$|$$
$$CH_2 - PO_3H_2$$

HEDP

1-Hydroxyethylidene-1,1-diphosphonic acid

$$OH$$
$$|$$
$$H_3C - C - PO_3H_2$$
$$|$$
$$PO_3H_2$$

EDTMP

Ethylenediaminetetra-(methylenephosphonic acid)

$$H_2O_3P - H_2C\diagdown \qquad \qquad \diagup CH_2 - PO_3H_2$$
$$N - CH_2 - CH_2 - N$$
$$H_2O_3P \quad H_2C\diagup \qquad \qquad \diagdown CH_2 \quad PO_3H_2$$

DTPMP

Diethylenetriaminepenta-(methylenephosphonic acid)

$$H_2O_3P - H_2C\diagdown \qquad \qquad \qquad \qquad \diagup CH_2 - PO_3H_2$$
$$N - CH_2 - CH_2 - N - CH_2 - CH_2 - N$$
$$H_2O_3P - H_2C\diagup \qquad | \qquad \diagdown CH_2 - PO_3H_2$$
$$CH_2$$
$$|$$
$$PO_3H_2$$

Figure 3.9 Constitution of organophosphonate sequestering agents. (*Source*: Ref. 123.)

the boiling treatment, alteration in count affected by both losses, and changes in tensile strength (generally an increase) [95]. But the most important characteristic of scoured fabrics is increased wettability, which is a necessary property for good uniform bleaching, dyeing, and finishing. Wettability must be obtained not only in capillary spaces between the fibers but also inside the fibers themselves (through absorption) [128]. While the scouring treatment is beneficial for wettability, when carried out under severe conditions, it induces fiber deterioration: e.g., the creation of crevices in fibers or dissolution of the cuticle and primary wall [129,130].

Since the scouring treatment contributes to the dissolution of a portion of shorter cellulosic chains, it bestows on cotton an average degree of polymerization higher than that of native cellulose. Otherwise, this alkali treatment, in contrast to mercerizing, does not induce profound changes in the fine structure of the fibers and has, for example, only a small effect on the degree of crystallinity of cotton [131, 132].

3.4 Technological Aspects of the Scouring Process

From the technological point of view, the scouring processes differ depending on whether they are carried out in a noncontinuous, semicontinuous, or continuous way.

Discontinuous Process

Kiering Process. The oldest scouring method is the discontinuous process and it is generally carried out under pressure in an autoclave called kier. This process is applicable to all types of goods: flock, hanks of yarns, and fabrics, and has potential for flock treatment of cotton (e.g., manufacture of cotton bandaging) and hanks of yarn for which scouring treatments and bleaching are carried out successively in the same circulating bath vessel. This method, however, is used less often for fabrics, its main competition being the more efficient semicontinuous and continuous processes. Kier scouring generally consists of the preliminary impregnation of fabric with an alkaline bath, then of its piling in a rope form in an iron or steel cylindrical vessel, and of circulation of the alkaline solution through the fabric while heating it.

Depending on its size, the scouring vessel is capable of containing between 250 kg and 5 tons of textile. The loading of the goods is carried out by mechanical piling so that it is as regular as possible. This piling regularity is necessary in order to avoid less packed areas associated with preferential bath circulation ("channeling") and treatment irregularities. After piling the preimpregnated goods, the scouring liquor is introduced from the bottom of the vessel to the desired volume in order to sweep out the air. To this effect a valve is left open until all air is displaced by steam. The heating of the bath takes place either by direct steaming, or by means of an external heater. The circulation of hot alkaline liquor through the fabric is carried out under pressure (around 130°C) during a period of 2-12 hr depending on the specific weight of fabric and the required scouring effect. After treatment, rinsing takes place by the introduction of hot water to dilute progressively the alkaline bath while avoiding the presence of air ("water-push rinsing"). Many designs of kiers have been put on the market [87,95], the chief types being the following:

Injector Kier. This is the simplest and oldest form of kier and consists of a cylindrical vertical vessel with a perforated false bottom. The lid is provided with two holes permitting the introduction and exit of the fabric. A steam injector at the base of a buffer pipe located in the center of the kier permits the heating of the bath and makes it circulate from bottom to top before spraying it on the fabric. The disadvantage of this kind of direct steam heating is the dilution of the bath by condensed steam.

Walsh's Kier. This design uses the more efficient centrifugal pump circulation in conjunction with an external multitubular heater. In this way, the dilution of alkaline liquor by steam is avoided. The bath circulates from the bottom of the vertical vessel to the heat exchanger and then is reintroduced in the vessel by means of a spreader.

Jefferson-Walker's Kier. This type is based on the principle of periodic withdrawal, heating, and return to the kier of successive charges of alkaline bath, the entire liquor being circulated every 10-12 min.

Gebauer's Kier. This consists of a vertical vessel provided with a perforated inner jacket. The alkaline solution, heated in an external exchanger, is injected at the top of the apparatus between the two concentric walls. There is horizontal or radial, rather than vertical, circulation of the liquor.

Matther and Platt's Kier. This is a horizontal kier which is charged with two wagons loaded with fabric. The wagons are filled outside the kier with cloth saturated with scouring liquor, which allows very regular piling. The wagons are provided with a perforated bottom and a telescopic connection to the circulating pipes at the bottom of the kier. The bath is heated by an external multitubular exchanger. The advantage of this kind of apparatus is that it allows the treatment of fabrics in rope form as well as in open width.

Jackson's Kier. This is an apparatus which is similar to a jig and permits open-width treatment. It is adapted for scouring heavy fabrics sensitive to crease marks [133]. This system consists of the winding of the saturated goods on a roller mounted on a special frame which goes into a horizontal kier. The fabric is wound from the first roller to a second one during the boiling process under pressure; the liquor is circulated by a pump and spread over the fabric as it passes from one roller to the other.

Jig Process. For small lots, for example of the order of 1000 m or less, scouring can be carried out in jigs. In general, the scouring is done at atmospheric pressure, although there are jigs which permit operating under increased pressure. The time of treatment is about 2-4 hr and the resulting scouring effect is only partial compared with that of treatment under pressure in a kier. In spite of a shorter treatment time, the contact between alkali-saturated fabric and air induces some cellulose degradation; that is the reason for which semi-immersed jigs have been developed. These allow for a decrease in the area of contact between the fabric and air but make winding without the formation of creases more difficult.

Semicontinuous Process. The semicontinuous process of scouring consists of the use of a pad-roll installation. After impregnation with the alkali bath, the fabric, in open-width form is quickly preheated in a steam chamber or by infrared radiation. When it has reached a high enough temperature, the fabric is introduced into a transportable reaction chamber heated by steam injection, where it will be stored on a rotating roller. An alternative to this procedure consists of using a fixed dwell chamber which is large enough to contain several cloth batches in roll form on movable carriages. The reaction time is on the order of 1-1.5 hr at 100°C. Although this process is semicontinuous, it can be considered as continuous if the installations composed of several batching devices in which the cloth content can reach 5000 m. The pad-roll installation gives good results only if the preheating of the fabric is controlled well enough that the temperature of the fabric is neither too high, causing a partial evaporation of the bath, nor too low to allow it to easily reach the necessary temperature inside the dwell chamber. The steam content inside this reaction chamber prevents the last layers of the cloth batch from cooling or drying.

A recent amelioration of the traditional pad-roll system consists of winding the fabric onto a rotating perforated roll placed in a treatment chamber provided with a liquid reception vat and a circulation pump [134]. The alkali treatment bath circulates through the material from the inside to the outside of the beam. The combination of the hydrodynamic pressure and the centrifugal force ensures the regularity of the bath circulation. This kind of installation, which was first developed by Heberlein and Jaeggli permits scouring, bleaching, and washing operations on the same fabric roll.

Continuous Process. The first continuous scouring process permitting high productivity consisted of treating the fabrics in rope form. The DuPont [114] and the Becco processes (developed by Buffalo Electro Chemical Company in 1936) are inspired by the Gantt piler, a device used as long ago as 1920. These processes are based on the use of a

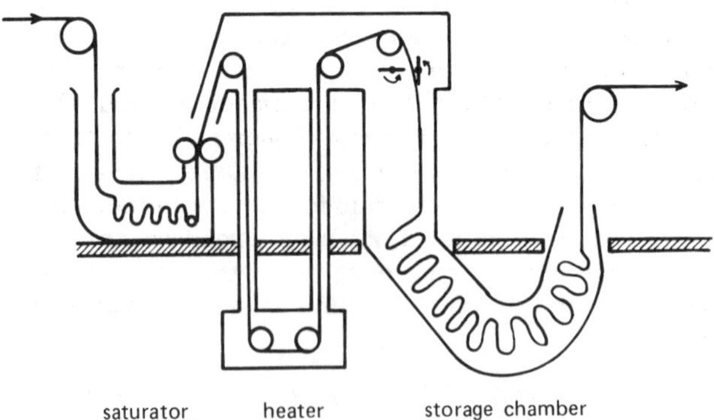

Figure 3.10 J box scouring process.

J-shaped fabric storage unit which is called J box. This system can be used for scouring as well as for bleaching with hydrogen peroxide, the two procedures taking place successively and continuously. After singeing, desizing, and rope washing, the cloth is impregnated by being passed through the scouring bath in a saturator. Upon leaving the saturator, a two-bowl rope squeezer ensures a constant and uniform liquor pickup. The cloth is then rapidly preheated by means of steam in a U-shaped heating tube, as seen in Fig. 3.10 (DuPont process). The fabric is stored (piled) in the insulated J box, where it can reside for periods up to 1 hr. In the case of the Becco system, there is no preheater and the cloth is directly heated in the J box, which is used as a steamer.

After the necessary period of time the rope leaves the J box and is washed in hot water. For a sodium hydroxide concentration of 40-60 g/liter and a maximum reaction temperature of 98°C, the production speed can reach 180 m/min, which implies a high storage capacity for the J box, taking into account the 30-60 min dwell time. So the capacity of the chamber must be on the order of 10,000 m of fabric. The large accumulation of cloth which is piled in the J box implies that this technique can be used only for fabrics which are not sensitive to creasing and bruising. For that reason it is desirable to turn to the open-width processes for the treatment of heavy polyester-cotton blend fabrics, which are increasingly common.

If we compare the two kinds of treatment, the advantages of rope processing over open-width processing are the lower chemical costs, the higher production speed, the suitability for a wide range of cloth widths without stopping the plant, and the efficiency and simplicity of the required equipment. But the advantages of open-width processing over rope processing are the suitability for fabrics sensitive to creasing, the better uniformity of treatment, the better warp tension control, and the lower filling distortion and surface abrasion [135].

Among the continuous open-width scouring methods, we must mention an ingenious system which permits a transformation of the pad-roll process in a completely continuous system [136]. This continuous batch or rebatching system is illustrated in Fig. 3.11. In the steaming box there are two winding stations, one above the other. The cloth is batched in two superimposed layers on the first winding roll; one layer comes from the other roll after partial aging and the second layer comes from the impregnation device. As soon as the batch is completely on the first winder, the direction of rotation is reversed and the second layer, which came first from the impregnation squeezer, is led to the second winder below the entering cloth, while the other layer, which has been completely aged, leaves the reaction chamber. For reaction periods of about 1-2 hr, this system allows fabric treatment with a production speed around 120 m/min. This system, nonetheless, has some inconveniences, such as being suitable only for the continuous

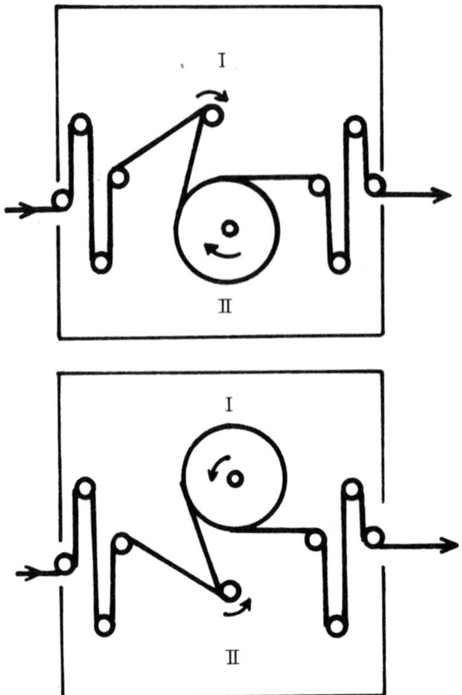

Figure 3.11 Continuous rebatching system.

treatment of fabrics with the same width; also, the risks of the formation of water marks (moiré) cannot be completely avoided in a batch process.

An alternative open-width continuous scouring process consists of using a roller steamer to prevent the risks of creasing and also of the formation of water marks (moiré). In this type of steamer system [87,135] the cloth is introduced continuously into a box where it is alternatively saturated with the sodium hydroxide solution and steamed. The passage of the fabric is ensured by guide rollers. The modern roller steamers no longer function by alternating the impregnation and steaming; they permit only a brief dwell time in the steamer after impregnation. This design is thus used preferably for the treatment of fabrics with a low seed and husk content. The dwell time is 2-3 min at 100°C after impregnation in a 60-100 g/liter sodium hydroxide solution. The rate of production is on the order of 60 m/min.

The use of roller steamers under pressure allows for some decrease in the steaming time and consequently in the space requirement for an equivalent production rate [115,119,135,137,138]. Such a pressure steamer is shown in Fig. 3.12. In this system, hydraulic sealing heads are subjected to a pressure which is slightly above the pressure inside the vessel where the cloth is guided on rollers. Each sealing head consists of two opposed polytetrafluoroethylene covered diaphragms and permits continuous cloth feeding and delivery. Another possibility consists of eliminating the guide rollers and depositing the fabric on a roller bed system, thus allowing the fabric to be in a relaxed state inside the pressure steamer, which is equipped with static lip seals [135]. The pressure steaming method enables a decrease in dwell time to 30-180 sec for a sodium hydroxide impregnation concentration equal to 60-100 g/liter and a temperature of about 130-140°C. In a vessel with a 60 m cloth content the production speed generally reaches 30-60 m/min.

For the purpose of permitting a longer continuous steaming time, another approach consists of an extrapolation of the J box system, using it for fabrics piled in open width. The quantity of stored fabric in the J- or U-shaped chamber must, however, be much lower compared to the treatment in rope form in order to avoid creasing. The result is reduced storage capacity, which necessitates a low production speed in order to maintain a reasonable dwell time. Another way of storing impregnated fabric consists of the use of a continuous pile storage steamer where the cloth is plaited down onto conveyor belts for the necessary time. This kind of conveyor steamer was first designed in the United States for chlorite bleaching [139] and was also suggested for scouring. In the open-width J box or the conveyor steamer, the accumulation of the goods is such that the risks of creasing are not suppressed.

It has been shown, however, that if the impregnated fabric was first guided in open width onto rollers for a time sufficient for swelling, it could subsequently be piled without risks of creasing [140-143]. This principle gave rise to the realization of a new kind of steamer,

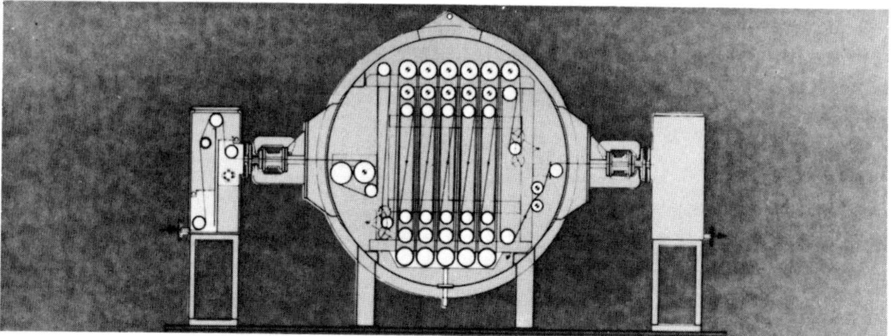

Figure 3.12 Continuous pressure steaming system. (Courtesy of Kleinewefers GmbH.)

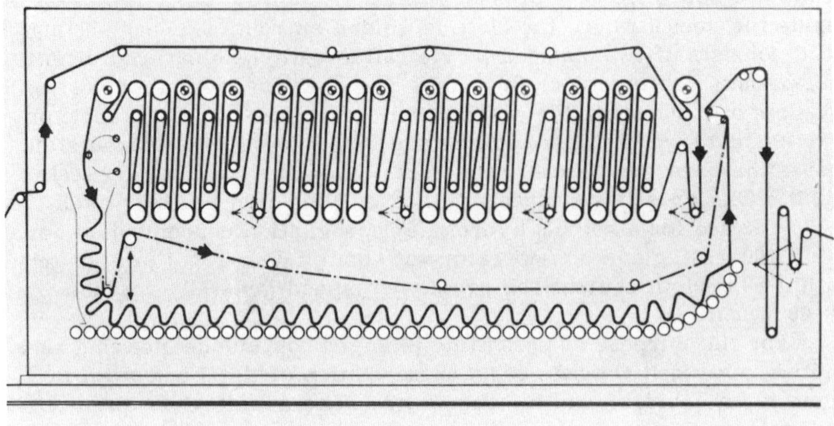

Figure 3.13 Roller bed steamer with preswelling zone. (Courtesy of Kleinewefers GmbH.)

an example of which is given in Fig. 3.13. This steamer combines a run on guide rollers and a subsequent deposit in folds. The fabric is preswollen in the tight strand roller section before being plaited down onto the roller bed for the required time. The roller bed conveyor can be replaced by a conveyor belt. This kind of installation permits 6-20 min dwell times at 100°C for sodium hydroxide concentrations around 30-60 g/liter and allows a production rate which can reach 100 m/min.

The relatively short dwell time in the continuous process, as compared to the noncontinuous processes, necessitates rapid wetting of the goods in order to obtain good and uniform alkali penetration. Instant wetting of the fibers to their core may be obtained by vacuum impregnation of the fabric with the sodium hydroxide solution before steaming. This technique can also be recommended for the wet-on-wet process [143,144].

3.5 Special Scouring Processes

Dewaxing with Solvent

Cotton waxes can be eliminated by approximate solvent treatment. For this purpose, the dry fabric is introduced into a bath of boiling solvent (trichlorethylene). After this treatment, the solvent is removed by means of boiling water or steam [145,146]. The fabric so treated possesses very little wax and is free from the solvent which has been exchanged with water. Dewaxed cotton still requires some forms of alkali treatment.

The solvent process eliminates the waxes but not the other cotton impurities. The remaining impurities can be removed simply by sodium carbonate treatment at atmospheric pressure. Emulsion scouring consists of treating the fabric with an emulsion containing an aqueous alkali solution emulsified in a chlorinated hydrocarbon which is then removed with steam and recovered [147].

Scouring of Colored Woven Goods

In the case of fabrics containing dyed yarns and gray cotton yarns, it is often necessary to scour and bleach the undyed part. Consequently, the dyes used must possess good fastness toward alkali treatments (for example, vat dyes and certain azoic dyes). The scouring bath has a strong reducing character because of the presence of the degradation products of the cotton impurities. The reducing character protects the cellulose from oxidation risks but it might cause the staining of leuco vat dyes and destruction of azoic dyes. It is therefore advisable not to add a reducing agent to the scouring bath but to incorporate a mild oxidizing agent such as sodium metanitrobenzenesulfonate. Furthermore, the scouring of color-woven goods is carried out preferably with sodium carbonate instead of sodium hydroxide.

Scouring of Blend Fabrics

In the case of polyester-cotton fabrics, whose usage is more and more common, it is necessary to take into account the sensitivity of polyester fibers to alkali solutions at the concentration and temperatures normally required for scouring. Thus it is advised to reduce the sodium hydroxide concentration to avoid the risk of polyester hydrolysis. The maximum allowable concentration of sodium hydroxide is a function of the temperature and of the length of treatment, taking into account the cotton proportion of the blended fabrics [148]. The use of certain scouring assistants can also be advised in order to guarantee the efficiency of the treatment while reducing the quantity of alkali [149].

Scouring Processes Combined with Other Pretreatment Steps

In the case of fabrics sized with polyvinyl alcohol, starch, or a mixture of the two, the addition of hydrogen peroxide to the sodium hydroxide solution allows for simultaneous desizing and scouring [150]. It is the same with the use of ammonium or sodium persulfate as a desizing agent incorporated in the alkaline scouring solution [151].

The scouring and bleaching processes can also be carried out simultaneously. In this case, the impregnation bath is composed of hydrogen peroxide, a stabilizer of hydrogen peroxide, sodium hydroxide, and a scouring assistant including a solvent and an emulsifier. This process, however, is only appropriate for fabrics with very few seeds and husks. The same kind of treatment bath, with a supplementary addition of ammonium persulfate, can be used to carry out simultaneously the operations of desizing, scouring, and bleaching as a pad-batch process at room temperature [152].

4. MERCERIZATION OF COTTON FIBERS

4.1 Introduction

In 1844 the English chemist Mercer wanted to reuse a sodium hydroxide solution conserved in a wooden barrel. To eliminate the wood particles he filtered this solution through cotton fabric and observed that the fabric shrank and became denser, and that its dyeing properties were improved; in 1850 he filed a patent describing this method of alkali treatment of cotton fabrics [23]. Forty years later, in 1890, Horace Lowe showed that by preventing the shrinkage of cotton fabric during alkali treatment, the fabric becomes more lustrous [153].

Currently, the alkali treatment of cotton materials is commonly used and we can make the distinction between treatment under tension (*tension mercerization*) according to Lowe and treatment without tension (*slack mercerization*) according to Mercer.

The mercerization process is carried out for the following purposes [154,155]:

- Improving luster and obtaining a silky look
- Improving the tensile strength
- Retaining a greater proportion of tensile strength after easy care finishing
- Obtaining improved dimensional stability
- Increasing uniformity of dyeing and improving color yield (resulting in savings in the cost of dyestuff)
- Improving elasticity and obtaining stretch materials (with slack mercerization)

Mercerization consists of impregnating the textile material with a concentrated solution of cold sodium hydroxide, keeping it in contact with this cold solution during a given time with or without tension, and subsequently rinsing it. The textile material can be present in the form of hanks of yarn or piece goods. Cotton materials may be mercerized in the gray state without any pretreatment, as well as after preliminary desizing, after scouring, or after scouring and bleaching. In the case of piece goods, mercerization can involve wet or dry fabrics.

The properties of mercerized cotton fibers, particularly the structural changes, have been studied by numerous authors for the purpose of clarifying the mechanism of the action of sodium hydroxide on cotton fibers and its practical consequences [156].

4.2 Influence of Mercerization on the Structure of Cellulose Fibers

Treating cellulose fibers in a highly concentrated alkali solution induces fiber shrinkage in width and length which is inversely proportional to the tension applied. This shrinkage corresponds to an increase in the volume of cellulosic materials, a phenomenon which is caused by swelling [157,158].

Swelling also induces changes in the cross section of fibers. A fiber becomes rounder as the treatment is carried out under increased tension. At the same time the fiber surface becomes more lustrous. The convolutions of the fibers disappear on swelling, possibly because the outer fibrillar layers of the fibers are more contracted than the inner layers; the result is the untwisting of the fibers [34,159].

The effect of mercerization is not, however, closely connected to swelling, but rather involves internal modification in the arrangement of the fibrils. The effect of mercerization takes place only at sodium hydroxide concentrations higher than 150 g/liter, where the swelling value is not the highest [7]. In fact, the range of alkali concentrations at which mercerization takes place is determined by the presence of solvated dipole hydrates, the size of which permits the solution to penetrate into the high-lateral-order portions of the fibers as well as into the more accessible parts. During this process the interchain forces are weakened, but the strength of the fibers is recovered by deswelling and drying. The transformation of native cellulose to mercerized cellulose is an irreversible exothermic phenomenon because of the modification of the crystalline network [160-162]. The fibrils do not lose their identity but change in dimension and orientation depending on the conditions of mercerization.

The most important effect of mercerization on the fine structure of cellulose fibers consists of changing the crystal lattice from cellulose I to cellulose II. The dimensions of the monoclinic unit cells of the two crystalline forms are given in Table 3.16 [163].

Table 3.16 Dimensions of Cellulose Crystalline Unit Cells

Crystalline form	a (Å)	b (fiber axis) (Å)	c (Å)	β (degrees)
Cellulose I	8.35	10.3	7.9	84
Cellulose II	8.14	10.3	9.14	62

Source: Ref. 163.

The conversion of the crystalline network results in a progressive rearrangement or shift of sheets of cellulose chains in crystallites of extended antiparallel chains [164,165]. The changes in crystalline structure produced by the mercerization process may occur during the swelling process, during the formation of soda cellulose, during the rinsing process (soda cellulose decomposition), and during drying [156]. If mercerization, for example, takes place by means of a 300 g/liter sodium hydroxide solution at 10°C, complex soda cellulose V forms (see Fig. 3.1). During rinsing soda cellulose V becomes soda cellulose I, then eventually soda cellulose IV, finally giving rise to a blend of cellulose I, cellulose II, and amorphous cellulose [10,166].

During the decomposition of soda cellulose I the temperature plays a very important role; less cellulose II is obtained as the rinsing temperature rises [166,167]. At the same time, when cellulose II appears we observe an increase in the proportion of disordered, accessible material [168-171]. This lateral-order decrease or decrystallization could be the step which imparts to the cellulose chains the mobility necessary for conversion from cellulose I to cellulose II [172].

In a first approximation, we can assume that there is an alteration of amorphous and crystalline parts in fibrous materials. From this hypothesis it is possible to define a crystallinity index or lateral-order index which can be computed by means of x-ray diffraction measurements [173-176]. Moreover, this method can be adapted to the case of mercerized cotton fibers which contain two distinct crystalline phases (cellulose I and cellulose II) [177,178]. The x-ray diffraction measurements allow for a precise determination of the changes occurring in the crystalline structure of cotton, depending on the conditions of mercerization. Thus the influence of sodium hydroxide concentration is illustrated in Fig. 3.14 in the case of cotton yarns mercerized for 60 sec at 20°C with constant length (0% stretch) [179]. It appears that cellulose II begins to form only at a sodium hydroxide concentration of around 125 g/liter, a concentration which corresponds with the start of the formation of soda cellulose I [180]. The cellulose II content reaches a maximum value at a 300 g/liter sodium hydroxide concentration; this maximum of conversion to cellulose II is connected to the presence of the solvated dipole hydrate $NaOH \cdot 7H_2O$ in the greatest proportion [7]. The conversion of the lattice structure from cellulose I to cellulose II depends on alkali concentration but not on the level of swelling; complete conversion could be obtained at alkali concentrations of 5 N or above, when cellulose yarns are treated without tension in solutions of sodium, lithium, or potassium hydroxides at 0°C [181].

The temperature of mercerization also has an influence on the crystalline structure of cotton, as shown in Table 3.17; the cellulose I content and the total crystallinity index decreases as the temperature increases, whereas the cellulose II content follows an opposite course.

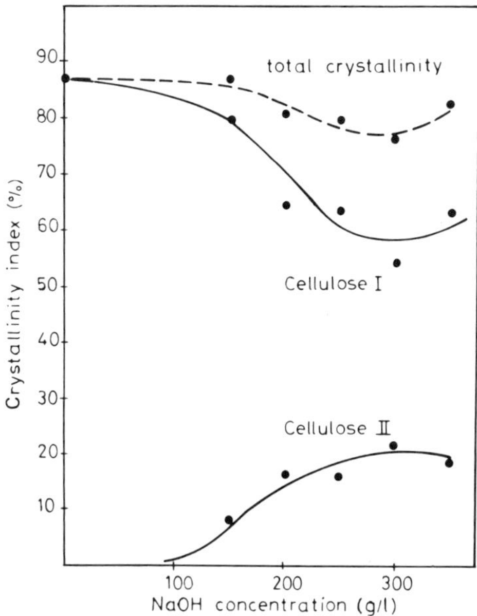

Figure 3.14 Influence of sodium hydroxide concentration on the crystalline structure of cellulose fibers. (*Source*: Ref. 179.)

Table 3.17 Influence of Mercerization Temperature on the Crystalline Structure of Cellulose Fibers (Sodium Hydroxide Concentration: 300 g/liter, with Constant Yarn Length)

Crystallinity indexes	Untreated sample	10°C	20°C	25°C
Cellulose I	81.0	58.4	53.7	47.1
Cellulose II	—	21.5	22.8	24.8
Total	81.0	79.9	76.5	71.9

Source: Ref. 179.

This phenomenon can be explained by the better penetration of sodium hydroxide hydrates in cotton yarns when the temperature increases. Different results are obtained according to whether the alkali treatment is carried out on elementary fibers, on fiber bundles, or (as in practice) on yarns or fabrics; thus, depending on the case, the homogeneity of alkali treatment will be different due to the fact that the swelling of the outer fibers in contact with the alkali solution will tend to decrease the sodium hydroxide penetration into the inner fraction of the material.

The tension applied to the cellulosic material, whether stretching or limiting the possibilities of shrinkage, plays a fundamental role in the transformations of the crystalline structure. The majority of changes in the fine structure due to mercerization is decreased by the application of increasing tension [156]. Fig. 3.15 shows the results obtained in the case of mercerization of yarns for 60 sec at 20°C in a 300 g/liter sodium hydroxide solution and for different values of the length variation during the process, i.e., stretch, constant length, or controlled shrinkage [179]. It appears that the ratio of conversion of cellulose I to cellulose II is more pronounced as the tension applied is lower. Thus, for increasing tensions soda cellulose I would transform preferentially into cellulose I during rinsing, the decrease of swelling of fibers under tension preventing the change of configuration into cellulose II [166]. On the other hand, total absence of tension (maximum level of shrinkage) would permit a total conversion into cellulose II [182,183]. From the point of view of crystallinity, the tension seems to play a less important role. The mercerization process also affects the size of crystallites, which can be estimated by measuring the leveling-off degree of polymerization (LODP) [169,183-186]. The method of measurement of the LODP value consists of carrying out hydrolysis of the cellulosic materials by means of 2.5 N hydrochloric acid. The degrees of polymerization of the residues are measured and the extrapolation at 100% residue represents the leveling-off degree of polymerization which corresponds to the dimensions of the crystallites, which are not attacked in an acid medium. The LODP value and consequently the size of crystallites decreases under the effect of mercerization, as seen in Table 3.18, and this decrease depends on the alkali concentration and on the tension applied to the textile material. The decrease of crystallite length is also influenced by the size of the alkali cation; thus the crystallite size is reduced to a greater extent when sodium hydroxide is replaced by potassium hydroxide, since the size of the cation is larger [161].

From the point of view of crystallite orientation relative to fiber axis, it is much more difficult to elucidate the effect of mercerization. Crystalline orientation can be computed by birefringence measurements, by x-ray diffraction measurement with azimuthal scanning, or by measurement of the sonic modulus. The measured values of the

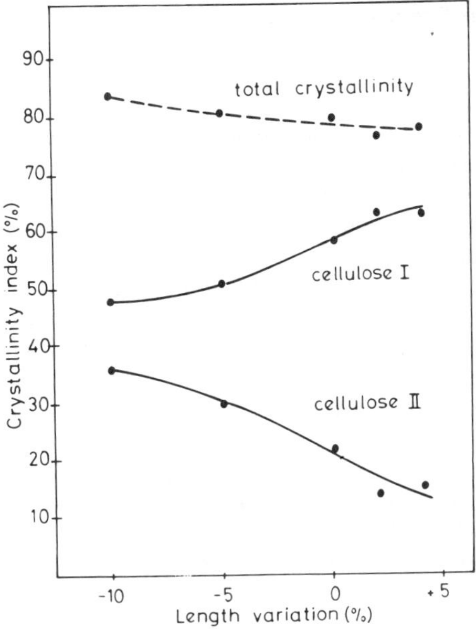

Figure 3.15 Influence of tension (length variation) on the crystalline structure of cellulose fibers. (*Source*: Ref. 179.)

orientation of cellulosic chains decrease in the case of slack mercerization and increase when the mercerization is carried out under tension [187-189]. However, the changes in orientation may arise from changes in the spiral angle of the fibrillar elements and not necessarily from changes in orientation of the crystallites within the fibrillar structure [190].

Table 3.18 Effect of the Mercerization Process on the Length of Crystallites

Type of material	LODP value
Native cotton	145-252
Slack-mercerized cotton	95-150
Tension-mercerized cotton	110-162

Source: Ref. 156,169.

4.3 Influence of the Mercerization Process on the Properties of Cotton Fibers

Luster

The luster of cellulosic materials (yarns and fabrics) depends on factors characteristic of the fiber (cross-sectional shape and its uniformity along the fiber, degree of convolution, surface roughness, and transparency) and of yarns (twist, regularity of structure, and fineness) [156]. Mercerization treatment induces an increase in luster of cotton fibers caused not only by the modification of their cross-section, which becomes circular under the effect of swelling, but also by the untwisting of the fibers, by the light scattering within the fiber (transparency), and by the modifications of the surface aspect of fibers.

Luster can be determined from reflectance measurements on cellulosic materials with incident light inclined at 45°; the reflected light is measured either perpendicularly to the surface of the material (diffuse reflectance) or viewed at 45° symmetrically to the incident light (specular reflectance). Luster is therefore defined by the contrast ratio of the specular to the diffuse reflectance [191]. The measurement of the specular reflectance can also be considered as an accurate evaluation of luster. Another method of luster characterization consists of image texture analysis. According to this method the fabric surface is photographed and the negative is scanned on a microdensitometer. Black spots on the negative, which correspond to glitter points on the fabric, are characterized in terms of intensity and area [192].

Luster increases as the tension applied to the fibers during mercerization is increased [191,193]. When mercerization is carried out without any tension (maximum shrinkage), no luster increase is observed. Although the fiber length is theoretically without importance on luster improvement, the presence of short fibers induces some decrease in fabric luster. These fibers are not submitted to tension, and it is important to eliminate them by singeing before mercerization. To retain an optimal level of luster, it is necessary not to release the tension before the sodium hydroxide concentration is lowered to about 60 g/liter. For this purpose, the first rinsing must be carried out under tension (stabilization phase).

The concentration of alkali in the mercerization bath affects the luster. It increases from 180 g/liter to a maximum at 260-300 g/liter, then decreases slightly [191]. Concerning the effect of temperature, the opinions are very divided.

Theoretically, luster increases with an increase in swelling and therefore with a decrease in the temperature of mercerization [194]. For that reason the mercerization bath is normally cooled down, but in a moderate way in order not to induce a superficial dissolution of the fibers, which would result in a decrease in luster [193]. However, recent experiments on yarns and fabrics have given a contradictory result, according to which luster increases with temperature increase

[195]. This result could be explained by the fact that in spite of lower swelling value, the penetration of sodium hydroxide into the fiber will be faster and more uniform at higher temperatures.

Mechanical Properties

Mercerization tends to increase the tenacity of cotton fibers. It has been suggested that the increase in fiber strength is related to the spiral angle and the orientation factor [196]. This explanation is, however, subject to controversy and it seems that orientation alone does not account for the difference in strength [197]. In fact, the increase in tenacity can also be attributed to a strengthening of the weak points along the fiber [183,198-200]. Among the weak points which are consolidated, we can mention the internal strains (spiral reversals), the places of low cross section area, and the places of high distortion in the macrostructure of cotton [156]. In addition to the modification of orientation and the consolidation of weak points, other important factors of strengthening could be the crystallite length and degree of crystallinity [201] as well as the removal of fractions of cellulose of very low degree of polymerization [156]. Increase in tenacity depends on the state of the material; experiments on fiber bundles resulted in significantly higher strength values than those obtained in the case of yarns and fabrics for which the mercerization degree was less homogeneous [202]. The elongation at break and the elastic modulus (Young's modulus) are also altered by the mercerization treatment. The tension applied to the cellulosic material has a great influence on the mechanical properties. Table 3.19 shows the results obtained in the case of mercerization of yarns for 60 sec at 20°C with 300 g/liter sodium hydroxide solution at several values of length

Table 3.19 Effect of Tension During Mercerization on the Mechanical Properties of Cotton Yarn

Length variation (%)	Tenacity ($cN \cdot tex^{-1}$)	Elongation at break (%)	Young's modulus ($cN \cdot tex^{-1} \cdot mm^{-1}$)
Unmercerized	12.9	7.7	1.9
-10	16.1	9.8	2.1
-5	16.6	6.7	3.1
0	17.5	5.1	4.1
+2	19.0	4.5	4.8
+4	19.3	3.9	5.2

Source: Refs. 179, 203.

variation (tension) during the process [179, 203]. The tenacity and the elastic modulus increases significantly when the mercerization tension increases. It is in connection with the degree of orientation of the fibers [204]. The extension at break, however, decreases when the tension increases and seems to be related to the variations of the degree of orientation [205]. If the mercerization is carried out without any tension (slack mercerization with a shrinkage of about 15-20%), the value of elongation at break is three times the value of the unmercerized material, thus conferring elastic properties to cotton fibers. This elasticity is the basis of the preparation of stretch materials with ability to extend and recover their initial length: after slack mercerization followed by easy care finishing, fabrics can posses 8-10% elastic elongation [206-209].

When the temperature of mercerization decreases, the increase in tenacity is more pronounced, especially when mercerization is carried out on single fibers or on bundles of fibers. If mercerization is carried out on yarns or on fabrics, the influence of temperature on the mechanical properties is much less significant, at least between 10 and 25°C [203]. The sodium hydroxide concentration of the mercerizing bath plays an important role as seen in Table 3.20. The tenacity and the Young's modulus increase with sodium hydroxide concentration reaches a maximum value as shown at 300 g/liter, which corresponds to the presence of the hydrate $NaOH \cdot 7H_2O$ in maximum proportion, imparting a better mercerization effect.

On the other hand, the elongation at break decreases to a minimum at the same alkali concentration and then increases again slightly. Above a sodium hydroxide concentration of 200 g/liter, the changes in the mechanical properties and in the tenacity are, however, of little significance (Table 3.20).

Table 3.20 Influence of Sodium Hydroxide Concentration on the Mechanical Properties of Cotton Yarn Mercerized at Constant Length

Sodium hydroxide concentration (g/l)	Tenacity ($cN \cdot tex^{-1}$)	Elongation at break (%)	Young's modulus ($cN \cdot tex^{-1} \cdot mm^{-1}$)
Unmercerized	12.9	7.7	1.9
150	15.6	6.3	2.8
200	17.1	5.7	3.6
250	17.3	5.3	3.9
300	17.5	5.1	4.1
350	17.4	5.4	3.8

Sources: Refs. 179, 203.

Dyeing Properties

Mercerization induces a decrease in the volume of crystalline region or an increase of the accessible area; moreover, this treatment modifies the texture of the fibers and principally their accessibility in an aqueous medium [179]. Consequently, mercerized fibers have greater moisture absorption capacity, are more reactive to aqueous chemical reagents, and are more accessible to dye molecules [210]. Mercerization increases the dye absorption (dye pickup), the rate of dyeing, and also the visual color yield — the mercerized material is of darker shade than the unmercerized when dyed with dye baths of same composition under the same conditions [211,212]. The improvement of the dyeing properties is especially marked when the mercerization treatment is carried out under decreasing tension. The dyeing properties vary only slightly with the treatment temperature and to a greater extent with the sodium hydroxide concentration. As an example, Table 3.21 shows the variations of the dye absorption at equilibrium and the changes in color yield as a function of sodium hydroxide concentration for yarns mercerized at constant length for 60 sec at 20°C [179]. In fact, the dyeing properties are optimal at sodium hydroxide concentrations of about 260-300 g/liter, but the gain observed between 170 and 300 g/liter does not justify the use of too high a concentration of alkali when we only intend to improve the dyeing properties without giving any importance to luster or to the strength

Table 3.21 Influence of Sodium Hydroxide Concentration on the Dyeing Properties of Cotton Yarn Mercerized at Constant Length (Dyeing with C.I. Direct Blue 1)

Sodium hydroxide concentration (g/l)	Dye absorption (w/w percent)	Gain in color yield relatively to unmercerized sample (%)
Unmercerized	1.48	–
150	1.52	+9
200	1.55	+15
250	1.58	+16
300	1.60	+26
350	1.58	+23

Source: Ref. 179.

of the material. In this case, the ideal treatment consists of slack mercerization with a 150-190 g/liter sodium hydroxide solution. This treatment is capable of decreasing dyeing unevenness resulting from differences in cotton maturity. Particular attention must be given to the drying of the mercerized goods because the dyeing properties are less improved if the drying is carried out at a higher temperature which induces some decrease in the accessibility of fibers.

The increase in color yield is such that it can allow for a saving of dyestuff on the order of 15-50%. Mercerized goods can show an apparently deeper shade than unmercerized goods, even when the percent of dye absorption is the same for both materials. The color yield has been defined by Kubelka and Munk as a function of the reflectance R of dyed material in terms of the ratio between a light absorption factor K and a light-scattering factor S, the factor K being in linear connection with the amount of absorbed dye C [213]:

$$\frac{K}{S} = \frac{(1 - R)^2}{2R} = \frac{aC}{S}$$

An increase in color yield is especially connected to a decrease in the light-scattering factor, this decrease being related to the change in cotton morphology; in effect, the more rounded form of the mercerized fibers is accompanied by a lower amount of internal scattering of the incident light [211, 212, 214, 215]. The color yield of mercerized goods can also be connected to the nonuniformity of the treatment. Thus, the mercerization of yarns and fabrics will almost inevitably give rise to a heterogeneous result. On entering the mercerization bath, the fibers on the surface of the yarns swell very quickly and give rise to a jamming effect at the periphery of each yarn. This phenomenon hinders the further penetration of the sodium hydroxide solution into the interior of the yarns, especially when the treatment is carried out under tension [216]. Since the level of mercerization of external fibers is higher than that of internal fibers, it follows that there is a preferential fixation of dyes on the fibers at the periphery of yarn. The penetration of the dyeing bath is also decreased due to the fact that the internal fibers of the yarn are compressed following the jamming effect caused by the swelling of the external fibers. The result of this heterogeneous treatment is a difference in the depth of dyeing between the periphery and the core of the yarns. If two yarns, one mercerized and the other unmercerized, are dyed with an equal amount of dye and if the mercerized yarn shows a preferential fixation of dye at its periphery, the apparent dye concentration of the mercerized yarn will be higher than that of the unmercerized yarn and there will be a higher color yield [179]. If we wish to avoid the peripheric dyeing of mercerized goods, it may be desirable to carry out the mercerization treatment at a higher temperature than commonly used. The increase in temperature induces a decrease in the degree of swelling, better penetration of the sodium hydroxide solution, and consequently better uniformity of mercerization and of subsequent dyeing [216].

Other Properties of Cotton Fabrics

In addition to luster and mechanical and dyeing properties, which are fundamentally the most important modifications related to the mercerization treatment, some other practical results are also obtained. Whether it is carried out with or without tension, mercerization of fabrics imparts a certain dimensional stability during the successive washing cycles. The dimensional stabilization is greater when the alkali treatment is carried out at a higher temperature [217].

Mercerization also gives a moderate improvement in crease recovery to cotton fabrics [218] as well as some protection against the decrease in tensile strength caused by easy care finishing.

4.4 Auxiliary Products for Mercerization: Wetting Agents

The contact time between goods to be mercerized and the alkali solution must be short (less than 60 sec) for productivity reasons. The success of the mercerization process depends also on the rate of the penetration of the viscous sodium hydroxide solution into the fibers of yarns or fabrics. In the case where goods to be mercerized are in the gray or nonscoured state, the hydrophobic waxes give the material a resistance to the penetration of the sodium hydroxide solution. For all these reasons, it is often advisable to include a wetting agent in the mercerizing liquor, which allows for a decrease in mercerization time and also ensures more uniform treatment.

The properties of wetting agents required for mercerization are as follows [219]:

Solubility in an alkaline medium
High wetting ability, particularly in a highly alkaline medium
Absence of affinity to the fiber (ease of elimination)
Efficiency at low concentration
Low foaming power
Stability under the conditions of sodium hydroxide recovery by centrifuge vacuum evaporation.

In the case of the mercerization of knitted fabrics which contain paraffin products, the wetting agent used must also possess dispersing power toward these products [220].

It is obvious that a single wetting agent cannot show all the properties listed above, and the choice of the required product very often requires a compromise which necessitates the use of a blend of wetting agents. The control of the efficiency of wetting agents in relation to their use in the mercerization process can be accomplished by measuring the rate of shrinkage of cotton yarns in sodium hydroxide solutions in the presence of these agents [220,221].

Normal wetting agents can be used for mercerization, but they must be associated with phenolic or cresolic products in order to maintain their properties in concentrated sodium hydroxide solutions. The presence of such products poses problems, however, from the point of view of water pollution because of their toxic and nonbiodegradable character [222, 223]. At present among the biodegradable wetting agents the most commonly used are certain alkylarylsulfonates, some organophosphonates, such as the sodium salt of methyloctylphosphonic acid [224]; alkylated diphenyloxide sulfates, such as sodium dodecyldiphenyloxide sulfate [225]; and, especially, sulfated aliphatic alcohols, such as 2-ethylhexyl sulfate [225]. The most efficient sulfated alcohols for the mercerization process are those with low molecular weight (chains with 4-8 carbons). Moreover, products with branched chains are more efficient than those with linear chains [219].

The addition of alcohols or diols to the wetting agent allows for an improvement in the penetration of the mercerizing liquor by diminishing the viscosity of the sodium hydroxide solution. The addition of high-boiling polyhydric alcohols enables the stabilization of the wetting agents made from sulfated alcohols while avoiding their precipitation during the process of recovering sodium hydroxide [219].

4.5 Hot Mercerization Process

Due to the rapid and extensive swelling of cotton fibers during their impregnation with cold caustic soda solution, the structure is compacted at the surface of mercerized goods (yarns or fabrics), and further penetration of sodium hydroxide is almost impossible. The result is a lack of uniformity when mercerization is normally practiced. Because of the lower level of swelling, mercerization of cotton yarns and fabrics at elevated temperatures may offer some advantages such as more even penetration of the sodium hydroxide solution into the inner parts of the yarns and consequently better uniformity of the alkali treatment [226, 227]. However, the conversion of the crystalline structure from cellulose I to cellulose II is retarded at high temperature, and photomicrographs of single fibers taken on hot-mercerized yarns reveal a skin-core appearance [228]. A reasonable approach is to use hot sodium hydroxide solution for a better penetration into the textile structure, and subsequently to allow the impregnated material to cool to an adequate temperature in order to have a good swelling value (i.e., a two-step process).

The process sequence of this two-step hot mercerization is as follows [216, 226, 227, 229-232]:

Saturation of the cotton material with sodium hydroxide solution, preferably under relaxed conditions, at a temperature between 50°C and the boiling point of the alkali solution. The immersion time ranges from 4-60 sec.

Controlled hot stretching following the saturation.
Cooling of the stretched material to a temperature less than 25°C.
Tension-controlled washing to a sodium hydroxide concentration of approximately 60 g/liter (stabilization).
Final washing under normal conditions.

Another possible method of hot mercerization consists of a wet-on-wet impregnation of the fabric immediately after washing at 95°C and a subsequent hot squeezing with high-speed steam injection. The heated fabric is impregnated with a sodium hydroxide solution at 30°C in the first step and at 20°C in the second. Thereafter, the stabilization sequence is normally carried out under tension-controlled conditions [223]. The advantages brought by the hot-mercerization process are numerous compared to the traditional process: most important, the degree of mercerization is more uniform. Due to the better penetration of the alkali solution, the desired effect of mercerization is achieved more rapidly. The period of contact between the cellulosic material and the sodium hydroxide solution can be reduced by approximately 50%, resulting in greater productivity or allowing a more compact mercerization unit. When the fabric is impregnated with the hot sodium hydroxide solution, it becomes highly plastic (but less elastic) and can be stretched to a greater degree than is possible in conventional mercerization. It is expected that higher tensile strength will be obtained because a higher proportion of the cellulosic material is modified [226, 227]. This expected increase in mechanical strength takes place only if the stretching process proceeds while cooling but, it is not observed if hot mercerization is completely carried out under isothermal conditions [232, 234]. The possibility of stretching the material to a high degree should induce improved luster, especially at higher temperatures [232, 234, 235]. Nonetheless, as could be noticed for the mechanical strength, the hot-mercerization process, when carried out without cooling during the stretching step, results only in luster similar to that obtained in the case of conventional mercerization [236].

Hot-mercerized cotton shows a much more uniform coloration than cold-mercerized cotton. The total dye uptake is better in the case of hot-mercerized cotton, but the color yield is less pronounced and decreases as the temperature of treatment increases, due to less peripheric dye absorption [216, 234]. Among other advantages, hot mercerization could also induce some improvement of easy care finishing characteristics (for example, a better wet crease recovery) and better receptivity to the subsequent operations of scouring and bleaching, which could consequently be carried out easier and faster [226].

The preceding considerations are contrary to the well-known opinion that a high level of mercerization corresponds to low temperatures of treatment. The results obtained by the two-step hot-mercerization process could, however, be explained as follows: During the impregnation with hot alkali solution (250-300 g/liter sodium hydroxide) only soda cellulose I can be formed. If it is accepted that the formation of soda cellulose V is necessary for obtaining the mercerizing effect, the transformation of Na-Cell I to Na-Cell V would take place during the cooling step, contrary to the notion that the phase boundary between soda cellulose I and soda cellulose V cannot be crossed in this direction [216].

While the hot-mercerization process appears to be advantageous in several respects, it is appropriate nevertheless to remember that cotton fibers impregnated with a concentrated sodium hydroxide solution will be subject to alkali degradation in the presence of air at high temperature.

Returning to more practical considerations, and in relation with a general view of the pretreatments of cotton fibers, hot mercerization can permit an elimination, or rather a simplification, of the scouring process by inserting a steaming step between hot alkali saturation and stabilization [226, 227, 230, 237]. This combined mercerization-scouring process can be carried out under the following conditions:

Hot saturation stage with concentrated alkali solution.
Steaming process immediately after saturation, the material being relaxed or stretched. The time of steaming can be on the order of 10 min at atmospheric pressure or can be reduced to 5 sec under pressure (flash scour step in pressure chamber at 130-140°C) [226, 227].
Cooling step and mercerization (chain or chainless stabilization).
Final washing.

Furthermore, the desizing process could be eliminated. The simplification of the pretreatment steps which takes place by the application of this combined process (i.e., a shorter process sequence and a reduction in the number of intermediate dryings) would induce consequently lower energy consumption and a cost saving between 30 and 40% [227].

4.6 Mercerization of Blend Fabrics

Sometimes it may be desirable to mercerize blend fabrics containing regenerated cellulose fibers as well as cotton fibers, in order to improve the dyeing properties or to diminish the difference in luster between the two kinds of fibers. However, the solubility in an alkaline medium of regenerated cellulose fibers, and in particular of viscose fibers, brings about losses in strength of the mercerized fibers. The dissolution of regenerated cellulose fibers takes place especially in dilute solutions of sodium hydroxide (around 80-130 g/liter), so that the real problem is not the mercerization step but the washing step, when the regenerated cellulose fibers become highly swollen and lose strength.

Even taking certain precautions, it is virtually impossible to mercerize pure viscose materials without inducing a high loss in strength. For such materials it is possible, however, to carry out an alkaline treatment without risk using a 50 g/liter sodium hydroxide solution, which is sufficient to obtain a significant improvement in the dyeing properties [238].

In the case of cotton-viscose blends, the following measures [45, 48, 58, 239] should be carried out:

Hot impregnation
Addition of electrolyte into the mercerizing liquor
Use of potassium hydroxide or a mixture of potassium and sodium hydroxide in order to reduce the solubilization of viscose fiber
Hot-water (90°C) rinse, the solubility of viscose fibers in an alkaline medium being less marked when the temperature increases
Addition of electrolyte (e.g., sodium chloride) in the rinsing bath, which diminishes the proportion of dissolved viscose

The first two measures cited diminish the effect of mercerization on the cotton part. It should be noted that potassium hydroxide is more expensive than sodium hydroxide. Precautions should be taken, especially at the rinsing step, to avoid the dumping of fibers and of yarns, caused by a coagulation phenomenon which occurs simultaneously with dissolution.

Table 3.22 shows the loss in weight and strength for fabrics containing different proportions of viscose treated on a chainless mercerizing range with a 300 g/liter sodium hydroxide solution then rinsed at 90°C [179]. The loss in strength increases with the increasing proportion of viscose in the blend fabrics. This decrease of the breaking strength is especially marked in the filling direction, since less tension is applied in this direction than in the warp direction.

Applied tension is consequently of prime importance from the point of view of the decrease of losses in strength of mercerized viscose fibers; this is also suggested by the results shown in Table 3.23, which concerns pure viscose yarns mercerized by fixing their length at well-defined values [179]. However, while the tension allows for a decrease of tenacity loss, it can be shown that rinsing at 90°C does not guarantee minimal mechanical degradation in the case of pure viscose materials. In reality, it is advisable only to mercerize blend fabrics which contain a certain proportion (at least 50% by weight) of cotton.

Polynosic fibers are more stable than ordinary viscose fibers in an alkaline medium [50,53,54,55]. These fibers, which have a high wet-elasticity modulus, retain this property after mercerization [54]. Cotton-polynosic blend fabrics can be mercerized without taking any special precautions [239]. It is advisable, however, to carry out the rinsing process with hot water [54, 240].

Table 3.22 Properties of Mercerized Cotton-Viscose Blend Fabrics (Chainless Mercerization, Washing at 90°C)

Type of fabric	Loss in weight (%)	Breaking load (daN)			
		Warp		Filling	
		Unmercerized	Mercerized	Unmercerized	Mercerized
Cotton (100%)	1.2	73.2	79.8	70.4	65.4
Cotton (85%), viscose (15%)	2.5	69.8	74.1	63.6	55.3
Cotton (67%), viscose (33%)	2.7	61.7	68.2	58.8	49.0
Cotton (50%), viscose (50%)	3.0	60.9	60.6	52.6	43.6
Viscose (100%)	6.3	50.6	49.2	40.8	29.5

Source: Ref. 179.

Table 3.23 Effect of Tension during Mercerization of Viscose Yarns on Tenacity (Sodium Hydroxide Concentration 300 g/l, Washing at 90°C)

Length variation (%)	Tenacity ($cN \cdot tex^{-1}$)
Unmercerized	16.8
-20	6.5
-10	8.2
-5	9.6
0	10.4
+2	11.4
+4	11.7

Source: Ref. 179.

HWM (high-wet-modulus) fibers show a high modulus of elasticity, which approaches that of ordinary viscose fibers when they are submitted to an alkali treatment [54]. The mercerization of goods containing HWM fibers induces a higher strength loss than that of polynosic fibers [53]. However, as in the case of viscose fibers, the mechanical degradation can be decreased by applying sufficient tension [51, 53]. HWM fibers are less hydrophilic than cotton fibers, and it might be possible to carry out the mercerization of cotton-HWM fiber blend fabrics in the absence of a wetting agent, the sodium hydroxide solution then being preferentially absorbed by the cotton component [241].

4.7 Technological Aspects of the Mercerization Process

Mercerization of Yarns

Yarns, essentially used as sewing threads or for embroidery, may be mercerized in the form of hanks. Hanks of yarn are wound onto two parallel rollers in the mercerizing machine, and these rollers are moved with a rotating movement. This rotation allows for a uniform distribution of alkali in the impregnation step within 60-90 sec of contact time. One of the two rollers is mobile and connected to a hydraulic system which enables it to control the stretching or shrinkage of yarns during the entire treatment. For controlling the tension applied to the yarns during the mercerization process, different cycles are recommended. For example, the yarn should be allowed to shrink on immersion in the alkali solution in order to facilitate impregnation, the tension being applied later [242]. Another possibility consists of applying a certain tension before impregnation, then increasing it gradually as the hanks come into contact with the alkali solution [243]. In fact, the stretching and relaxation cycles during impregnation are the subject of an often empirical approach and dependent on the material to be mercerized. In all cases, the stretching of yarns is carried out in such a way as not to exceed an elongation of 3-5% with reference to the initial length. A higher elongation allows better luster, but induces on the other hand a decrease in strength. After contact with the concentrated alkali solution, the hanks are rinsed, often with hot water, without any release in tension before the dilution of the sodium hydroxide solution reaches around 60 g/liter.

At this phase the hanks can be subsequently submitted to a scouring treatment when mercerized in the gray state. They undergo a final washing cycle, neutralization, wringing out, and drying [243,244]; the final step, drying, is of great importance from the point of view of dyeing uniformity [245]. Hank mercerization allows a higher luster than piece mercerization, due to the fact that the tension is more effectively applied to yarns than to fabrics because of crimp.

The mercerization of yarns can also be carried out from parallel threads wound onto cylindrical drums (warp mercerization), this process being most commonly used in the United States [242,243]. Since the winding into hanks and unwinding of mercerized hanks constitute expensive additional operations, working continuously from bobbin to bobbin has been suggested. One of the possibilities of continuous mercerization consists of winding the yarn impregnated with a sodium hydroxide solution onto two rotating cylinders whose axis form a certain angle between them, allowing a gradual stretching. The two cylinders can also be conceived with a profile adjusted in such a way that they enable the stretching or the relaxation of yarns during impregnation and washing-off cycles. These processes of continuous mercerization of yarns would permit reaching production speeds on the order of 400-600 m/min.

Mercerization of Woven Fabrics

Mercerization of woven fabrics is usually carried out continuously on mercerizing machines, sometimes equipped with stretching chains. In the case of the chain mercerizer the cotton fabric is impregnated with two alkali baths followed by squeezing devices. Between the two impregnation steps, the fabric threads around a set of timing or swelling drums which are driven only by fabric movement. The two squeezing steps, being carried out at different speeds, maintain the fabric under tension in the warp direction; the second impregnation can be followed by another air passage on drums. After impregnation, the fabric is fed into the clips of a tenter frame where it can be stretched in the filling direction. At the end of the run on the clip tenter, the fabric is rinsed by spraying with water (hot or cold), always being maintained under tension, in order to wash out the caustic soda to a concentration around 60 g/liter [246]. After the first rinse under tension, the fabric is completely rinsed and neutralized conventionally in an open-width washing machine.

In the case of the chainless mercerizing machine, the cloth is impregnated, squeezed, and passed during the first rinsing stage (stabilization step) into an expander unit with bowed rollers formed of cast-iron wheels fitted one into the other. These bowed rollers enable the stretching of the material in width, the filling extension being determined by the tension in the warp direction, a tension which is obtained by means of oscillating rollers. This kind of machine is less cumbersome than a chain mercerizing range, but the tension applied to the fabric, especially in the filling direction, is not as high and the level of luster obtained is consequently lower. On the other hand, the risks of fabric tears near the clips are avoided [239]. Moreover, the chainless machine offers the possibility of working with two widths of fabric, side by side or one on top of the other, thus increasing the production [247].

Mercerization of fabrics can also be carried out in a discontinuous way for small and medium lots. After impregnation and extension, the cloth can be wound onto a roll, the stabilization step taking place by spraying water during the unwinding of the first roll onto a second [248]. Another possibility consists of saturating the fabric with the sodium hydroxide solution on a movable chainless mercerizing carriage. After impregnation, the fabric is wound onto a perforated batching drum under uniform tension (see Fig. 3.16). The mercerizing carriage, moving on rails, retreats from the drum as the batch diameter increases. At the same time, water is continuously sprayed on the fabric being batched (stabilization step). After complete winding of the fabric, the carriage is removed and the washing process is carried out by pumping water through the roll of fabric, an operation which is intensified by the rotary movement of the drums [249-251].

The mercerization process can be carried out at various stages of fabric pretreatment: sized gray, desized gray, scoured, or more rarely bleached. Preliminary singeing is advised if luster improvement is of great importance. When mercerization is carried out after desizing and scouring, the rate of wetting of the fabric is improved and the contamination of the alkali solution with sizing products and impurities of cotton is reduced. If mercerization is carried out prior to scouring, the first rinsing bath coming from the stabilization section can be recovered and reused for the subsequent scouring process. Mercerization in the gray state has the advantage of not diluting the

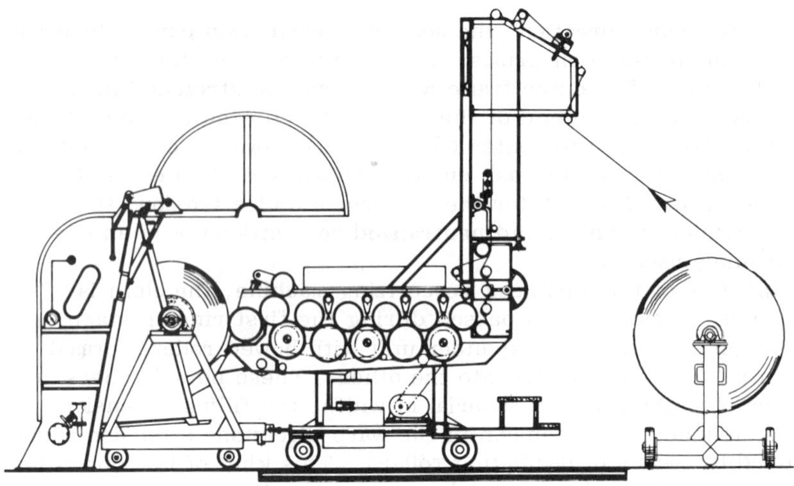

Figure 3.16 Mercerizing noncontinuous batching system. (Courtesy of Kleinewefers GmbH.)

sodium hydroxide solution as when we operate after desizing or scouring according to a wet-on-wet process.

The high concentration of sodium hydroxide in mercerization baths poses wastewater pollution problems. It is consequently possible to recover the alkali by vacuum evaporation of the rinsing baths [252, 253]. This technique is, however, virtually limited to the mills which carry out mercerization after desizing and scouring, since the sodium hydroxide is soiled when gray fabric is mercerized.

For mercerization in the gray state, for reducing the contact time of the fabric with the concentrated alkali, it is possible to use a vacuum impregnation device [251-254]. This method, also named the core-mercerizing process, is based on the use of a vacuum cap or vacuum chamber which encloses a section of the first roll of the impregnation tank (see Fig. 3.17). The fabric is guided through the vacuum chamber while the air is removed down to 80 torr pressure. On leaving the vacuum chamber, the fabric is immersed immediately into the sodium hydroxide solution, the lower part of the vacuum cap being placed beneath the level of this solution. This method will permit a decrease of the alkali dwell time and, while not requiring the use of any wetting agent, will guarantee better penetration of the alkali solution into the core of the fabric. If luster and strength are not required, slack mercerization may be carried out with only the purpose of improving the dyeing properties or obtaining stretch characteristics. In this case, the fabric is impregnated, squeezed, and then wound onto a roll to be stored for about 1 hr before the rinsing operations. This treatment which allows for a maximum value of shrinkage can also

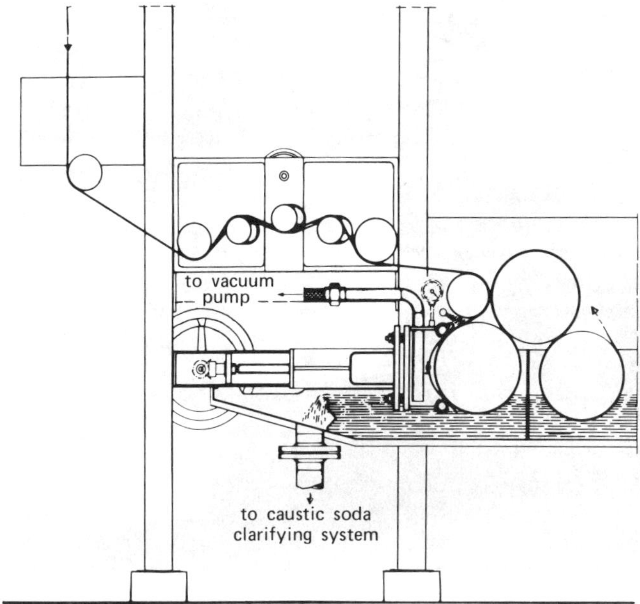

Figure 3.17 Vacuum impregnation system. (Courtesy of Kleinewefers GmbH.)

be carried out on a jig. Sometimes the fabric is allowed to shrink in one direction only. In this case, the impregnation is first made with shrinkage in both directions, and the fabric is subsequently stretched in the warp or the filling direction during the stabilization step.

Mercerization of Knitted Fabrics

For reasons similar to those of woven fabric mercerization, it would be desirable to submit knitted fabrics to an alkali treatment under tension. However, knitted fabrics are easily distorted and extended under relatively heavy loads; during the alkali treatment, these distortions become permanent. It is difficult to avoid this problem [255]. It is possible to work in a noncontinuous way by rinsing the knitted fabrics after impregnation and winding onto a perforated beam [255] or by carrying out the stabilization by water spraying during the batching after impregnation [256,257].

In order to use a chainless mercerizing range for the treatment of knitted fabrics, it is necessary to observe certain conditions [252, 255,258]:

Use of a special entry system for knitted fabrics, situated as close as possible to the zone of impregnation and guaranteeing a guidance carefully controlled in length and width

(a)

(b)

Figure 3.18 Mercerization process for tubular knit goods. [(a) Courtesy of Lindauer Dornier GmbH.] [(b) Courtesy of Thies GmbH.]

Positive driving of the lower rollers during impregnation and stabilization
Control of the longitudinal tension at short intervals, the machine being separated into several compartments, each with separate driving control units

Thus, the applied tension can be controlled more easily in relation to the state of the material at the various treatment phases (impregnation and stabilization).

It is also possible to associate a preliminary stretching on a tenter frame to the chainless mercerizing machine [258, 259]. The knitted fabric is thus put at a definite width on the clip tenter before its immersion in the caustic soda solution, and the stabilization process is carried out next in the chainless compartment.

The mercerization of tubular knitted fabrics in a flat state poses certain problems because creases are set in at the edges and cannot easily be removed in subsequent processings [255]. An ingenious solution to these problems is illustrated in Fig. 3.18 [260-262]. After impregnation, the knitted fabric in tubular form is stabilized over expander circular rings which form a vertical floating tower or stabilizing cigar. Thus, the dwell time with concentrated alkali solution and the first rinsing step (stabilization) may be carried out under regular tension along the whole circumference of the tubular knit, guaranteeing the absence of selvage marks.

REFERENCES

1. R. Bartunek, *Das Papier 9*, 254 (1955).
2. R. Bartunek, *Das Papier 16*, 568 (1962).
3. G. E. Collins, *J. Text. Inst. 16*, T123 (1925).
4. W. F. Du Bois, *Melliand Textilber. 46*, 837 (1965).
5. S. U. Pickering, *J. Chem. Soc. (London) T63*, 890 (1893).
6. R. Bartunek, *Kolloid Z. 133*, 125 (1953).
7. R. Bartunek, *Kolloid Z. 146*, 35 (1956).
8. R. Bartunek, *Das Papier 7*, 153 (1953).
9. R. Bartunek, *Das Papier 6*, 356 (1952).
10. H. Sobue, H. Kiessig, and K. Hess, *Z. Phys. Chem. B43*, 309 (1939).
11. K. Hess and C. Trogus, *Z. Phys. Chem. B11*, 381 (1931).
12. J. B. Calkin, *J. Phys. Chem. 40*, 27 (1936).
13. C. Trogus and K. Hess, *Cellulosechem. 15*, 1 (1934).
14. K. Hess and C. Trogus, *Z. Elektrochem. 42*, 696 (1936).
15. K. Hess and J. Gundermann, *Ber. Dtsch. Chem. Ges. 70*, 529 (1937).
16. W. Schramek, *Kolloidbeihefte 40*, 87 (1934).
17. J. Chedin and A. Marsaudon, *Makromol. Chem. 15*, 115 (1955); *20*, 57 (1956); *33*, 195 (1959).

18. I. Sakurada and S. Okamura, *Kolloid Z. 81*, 199 (1937).
19. K. Lauer, *Makromol. Chem.* 7, 5 (1951).
20. P. H. Hermans and A. Weidinger, *J. Colloid Sci. 1*, 185 (1946).
21. C. Trogus and K. Hess, *Z. Elektrochem. 42*, 710 (1936).
22. J. R. Katz and W. Vieweg, *Z. Elektrochem. 31*, 157 (1925).
23. J. Mercer, British Patent 1396 (1850).
24. T. B. Bright, *J. Text. Inst. 17*, T396 (1926).
25. A. T. Moore, L. W. Scott, I. V. De Gruy, and M. L. Rollins, *Text. Res. J. 20*, 620 (1950).
26. *ASTM Standards on Textile Materials, D 1442-1454 T*, p. 336, Philadelphia (1954).
27. P. A. Roelofsen, *Encylopaedia of Plant Anatomy* Vol. 3, Pt. 4, Nikolassee, Berlin (1959).
28. D. Bechter and D. Fiebig, *Melliand Textilber.* 57, 573 (1976).
29. E. Heuser and R. Bartunek, *Cellulosechem.* 6, 19 (1925).
30. W. Weltzien, *Textilber.* 7, 338 (1926).
31. G. Saito, *Kolloidbeihefte 49*, 365 (1939).
32. J. D'Ans and A. Jaeger, *Cellulosechem.* 6, 137 (1925).
33. J. O. Warwicker, *J. Appl. Polym. Sci. 13*, 41 (1969).
34. C. R. Nodder, *J. Text. Inst. 13*, T161 (1922).
35. J. H. Carra, V. W. Tripp, and R. S. Orr, *Text. Res. J. 32*, 1041 (1962).
36. B. R. Porter and R. S. Orr, *Text. Res. J. 35*, 159 (1965).
37. S. K. Raman and H. K. Rouette, *J. Text. Inst. 61*, 412 (1970).
38. D. Bechter and D. Fiebig, *Text. Praxis 25*, 738 (1970).
39. D. Bechter and D. Fiebig, *Text. Praxis 34*, 531 (1979).
40. D. Bechter and E. Schollmeyer, *Text. Praxis 34*, 1245 (1979).
41. H. Rath, *Lehrbuch Der Textilchemie* (H. Rath, Ed.), Springer Verlag, Berlin, p. 197 (1963).
42. G. Melzer, *Textil-und Faserstofftechnik 3*, 371 (1953).
43. H. Rath, *Lehrbuch der Textilchemie* (H. Rath, Ed.), Springer Verlag, Berlin, p. 21 (1963).
44. G. F. Davidson, *J. Text. Inst. 27*, T112 (1936).
45. O. Mecheels, *Melliand Textilber. 17*, 725, 804 (1936).
46. H. Hampe, B. Philipp, and J. Baudisch, *Faserforsch. und Textiltechn. 23*, 425 (1972); *24*, 113 (1973).
47. R. H. Peters, *Textile Chemistry*, Vol. 2, (R. H. Peters, Ed.), Elsevier, Amsterdam, p. 364 (1967).
48. H. Rath, *Z. Ges. Textilind. 39*, 357, 371, 393 (1936).
49. H. Korte, E. Kayser, and W. Waibel, *Melliand Textilber. 17*, 801, 864 (1936).
50. G. Centola, *Melliand Textilber. 44*, 551 (1963).
51. G. Centola and F. Riva, *Ric. Doc. Tessile 2*, 1 (1965).
52. J. Chabert and M. Peter, *Bull. Scient. ITF 24*, 419 (1970).
53. L. Szebo, *Faserforsch. und Textiltechn. 21*, 422 (1970).
54. F. Girard, *Chemiefasern 20*, 884 (1970).
55. P. Bouriot and F. Delestang, *Bull. Scient. ITF* 5, 1 (1976).

56. G. F. Davidson, *J. Text. Inst. 25*, T174 (1934).
57. G. F. Davidson, *J. Text. Inst. 23*, T95 (1932).
58. I. Rusznák, R. Szentpaly, and J. Kovacs, *Melliand Textilber. 43*, 1208 (1962).
59. M. Lewin, *Text. Res. J. 35*, 979 (1965).
60. M. Lewin, and L. G. Roldan, *Text. Res. J. 45*, 308 (1975).
61. M. Albeck, A. Ben-Bassat, and M. Lewin, *Text. Res. J. 35*, 935 (1965).
62. O. Samuelson and A. Wennerblom, *Svensk. Papperstidn. 57*, 827 (1954).
63. W. M. Corbett and J. Kenner, *J. Chem. Soc.* p. 1431 (1955).
64. G. Machell and G. N. Richards, *J. Chem. Soc.* p. 4500 (1957).
65. H. R. Mauersberger and E. W. K. Schwartz, *Rayon and Staple Fiber Handbook*, 3rd ed., Rayon Handbook Co., New York, p. 5 (1939).
66. E. Entwistle, E. H. Cole, and N. S. Wooding, *Text. Res. J. 19*, 527, 609 (1949).
67. H. Staudinger and J. Urisch, *Zellstoff Papier 18*, 690 (1938).
68. J. F. Haskins and M. J. Hogsed, *J. Org. Chem. 15*, 1264 (1950).
69. B. Warner, *Makromol. Chem. 11*, 87 (1953).
70. F. Alvang and O. Samuelson, *Svensk. Papperstidn. 60*, 31 (1957).
71. H. G. Muller, *Textil Rundschau 12*, 622, 671 (1957); *13*, 67 (1958).
72. G. V. Schultz and F. Mertes, *Das Papier 13*, 469 (1959).
73. T. N. Kleinert, *Textil Rundschau 15*, 55 (1960).
74. D. M. MacDonald, *Tappi 48*, 708 (1965).
75. G. N. Richards, *Cellulose and Cellulose Derivatives*, Vol. 5 (N. M. Bikales, Ed.), Wiley Interscience, New York, p. 1007 (1971).
76. J. A. Mattor, *Tappi 46*, 586 (1963).
77. J. A. Mattor, A Study of the mechanism of alkali cellulose autooxidation, Thesis, Appleton, Wis. (1963).
78. H. G. Müller, Über den Abbau von alkalisierte Cellulose mit molekularen Sauerstoff, EMPA, Bericht no. 186, St. Gall, Switzerland (1955).
79. A. Kantouch and J. Meybeck, *Bull. Scient. ITF, V 23*, 189 (1969).
80. R. H. Peters, *Textile Chemistry*, Vol. 2 (R. H. Peters, Ed.), Elsevier, Amsterdam, p. 177 (1967).
81. R. Freytag, *Textilveredlung 11*, 82 (1976).
82. G. F. Goldwaith and J. D. Guthrie, *Matthew's Textile Fibers* (H. R. Mauersberger, Ed.), Wiley Interscience, New York (1954).
83. A. J. Turner, *J. Text. Inst. 40*, T973 (1949).
84. E. R. McCall and J. F. Jürgens, *Text. Res. J. 21*, 19 (1951).

85. E. Frieser, *Textil Rundschau 15*, 472, 591 (1960); *16*, 76 (1961).
86. G. V. Hornuff and H. Richter, *Faserforsch. u. Textiltechn. 15*, 115, 165 (1964).
87. J. T. Marsh, *An Introduction to Textile Bleaching* (J. T. Marsh, Ed.), Chapman and Hall, London, p. 167 (1946).
88. S. Zeisel in *Rohstoffe des Pflanzenreiches*, Vol. 4, (J. Wiesner, Ed.), Leipzig (1926).
89. P. B. Marsh, H. D. Baker, T. Kerr, and M. L. Butler, *Text. Res. J. 20*, 288 (1950).
90. V. W. Tripp, A. T. Moore, and M. L. Rollins, *Text. Res. J. 21*, 886 (1951).
91. J. H. Kettering, C. F. Goldwaith, and R. M. Kraemer, *Text. Res. J. 16*, 627 (1946).
92. L. V. Lecomber and M. E. Probert, *J. Text. Inst. 16*, T338 (1925).
93. P. H. Clifford and M. E. Probert, *J. Text. Inst. 15*, T401 (1924).
94. R. G. Fargher and L. Higginbotham, *J. Text. Inst. 15*, T419 (1924); *17*, T233 (1926); *18*, T283 (1927).
95. S. R. Trotman and E. L. Thorp, *The Principles of Bleaching and Finishing of Cotton* (S. R. Trotman and E. L. Thorp, Eds.), Griffin, London pp. 75-302 (1927).
96. A. E. Bailey, *Cottonseed* (A. E. Bailey, Ed.), Interscience, New York p. 472 (1948).
97. G. Knecht and G. Allan, *JSDC 25*, 142 (1911).
98. A. C. Chibnall, S. H. Piper, A. Pollard, E. F. Williams, and P. N. Sakai, *Biochem. J. 28*, 2189 (1934).
99. C. Zerbe, in *Mineralöle und verwandte Produkte*, Springer Verlag, Berlin, 1952.
100. A. Kling and H. H. Hofstetter, *Seifen-Öle-Fette-Wachse 92*, 323 (1966).
101. K. Merkenich and A. Kling, *Textil Praxis Int. 34*, 1392, 1497 (1979).
102. K. Hess, *Melliand Textilber. 24*, 334 (1943).
103. M. M. Tschilikin, *Melliand Textilber. 10*, 883 (1929).
104. M. Freiberger, *Melliand Textilber. 14*, 302 (1933).
105. R. G. Fargher and J. C. Withers, *J. Text. Inst. 13*, T1 (1922).
106. L. Friedman and O. L. Kline, *J. Biol. Chemistry 184*, 599 (1950).
107. L. F. Cavalieri and M. L. Wolfrom, *J. Amer. Chem. Soc. 68*, 2022 (1946).
108. A. C. Walker and M. H. Quell, *J. Text. Inst. 24*, T131 (1933).
109. R. L. Whistler, A. R. Martin and M. Harris, *ADR 13*, 252 (1940); (1940); *Text. Res. J. 10*, 269 (1940).
110. A. Geake, *J. Text. Inst. 15*, T81 (1924).

111. AATCC Technical Manual 56, Test Method 39-1980, p. 286 (1980).
112. W. A. S. White, H. J. Ross, and N. F. Crowder, *J. Text. Inst. 50*, P3 (1959).
113. A. J. Pennings and W. Prins, *J. Polym. Sci. 58*, 229 (1962).
114. D. J. Campbell, *ADR 33*, 293 (1944).
115. P. Ney, *Textil Praxis 19*, 707 (1964).
116. B. K. Easton, *ADR 37*, 985 (1964).
117. G. Richter, *Textil Praxis 20*, 897 (1965).
118. A. Lehn, *Textilveredlung 1*, 257 (1966).
119. C. Duckworth, J. V. Horsley, and J. J. Thwaites, *JSDC 88*, 281 (1972).
120. J. Diemunsch and R. Freytag, *Bull. Scient. ITF, V 5*, 37 (1973).
121. H. Grunert, *Textil Praxis Int. 34*, 1188 (1979).
122. R. Freytag, *Colourage 22*, 35 (1975).
123. X. Kowalski, *J. AATCC 10*, 161 (1978).
124. R. Blanckenhorn, *Melliand Textilber. 50*, 1063 (1970).
125. D. Bassing and J. Reiss, *Dtsch. Färber Kalender 79*, 104 (1975).
126. G. Weisse, *Textiltechnik 27*, 494 (1977).
127. A Kling, *Dtsch. Färber Kalender 80*, 35 (1976).
128. J. J. De Boer, *Text. Res. J. 50*, 624 (1980).
129. H. U. Mehta, P. Neelakantan, and J. T. Sparrow, *Indian J. Text. Res. 2*, 24 (1977).
130. G. M. Venkatesh, *Text. Res. J. 49*, 75 (1979).
131. K. Schwertassek and K. Hajek, *Faserforsch. u. Textiltech. 3*, 349 (1952).
132. L. Segal, J. J. Creely, A. E. Martin, Jr., and C. M. Conrad, *Text. Res. J. 29*, 786 (1959).
133. C. L. Jackson and E. W. Hunt, German Patent 127,002 (1902).
134. D. Angstmann and K. H. Rucker, *Textil Praxis Int. 28*, 464 (1973).
135. J. T. Bowden, *Textilveredlung 8*, 173 (1973).
136. P. Buschmann, *Melliand Textilber. 35*, 304 (1954).
137. Matther and Platt Ltd., *Inter. Dyer 145*, 206 (1971).
138. J. J. Thwaites, *Indian Text. J. 84*, 159 (1973).
139. Mathieson Alkali Corp., *ADR 33*, 536 (1944).
140. C. A. Theusink, *Dtsch. Färber Kalender 81*, 102, 108 (1977).
141. H. Weber, *Textilveredlung 13*, 215 (1978).
142. G. Schiffer, *L'Industrie Textile 1087*, 245 (1979).
143. G. Schiffer and B. D. Bähr, *Textil Praxis Int. 33*, 167 (1978).
144. G. Schiffer, *Inter. Dyer 159*, 463 (1978).
145. W. A. S. White, *Textil Rundschau 7*, 2 (1964).
146. M. D. Dixit, *Colourage 22*, 28 (1975).
147. I.C.I. Americas, Inc., U.S. Patent 4,076,500 (28 February 1978).

148. K. H. Rucker, *Textil Praxis 28*, 274 (1973).
149. W. Präger, *ADR 67*, 24, 49 (1978).
150. M. H. Rowe, *J. AATCC 10*, 215 (1978).
151. H. Carron, *Teintex 42*, 459 (1977).
152. H. Carron, *Teintex 45*, 3 (1980).
153. H. Lowe, British Patent 4452 (1890).
154. S. A. Heap, *Intern. Dyer 154*, 374 (1975).
155. L. A. Stiver, *AATCC National Technical Conference Proceedings*, p. 77 (1977).
156. J. O. Warwicker, R. Jeffries, R. L. Cobran, and R. N. Robinson, Shirley Institute Pamphlet no. 93, Manchester (1966).
157. H. F. Coward and L. Spencer, *J. Text. Inst. 14*, T32 (1923).
158. R. W. Willows and A. C. Alexander, *J. Text. Inst. 13*, T237 (1922).
159. C. Steinbrinck, *Biol. Zbl. 26*, 657 (1906).
160. G. E. Collins and A. M. Williams, *J. Text. Inst. 14*, T287 (1923).
161. S. M. Neale, *J. Text. Inst. 20*, T373 (1929).
162. G. Lal, *Text. Res. J. 44*, 313 (1974).
163. K. R. Andress, *Z Phys. Chem. B4*, 190 (1929).
164. F. J. Kolpak and J. Blackwell, *Polymer 19*, 132 (1978).
165. M. Takai and J. R. Colvin, *J. Polym. Sci., Polymer Chem. Ed. 16*, 1335 (1978).
166. R. Jeffries and J. O. Warwicker, *Text. Res. J. 39*, 548 (1969).
167. J. Hayashi, T. Yamada, and K. Kimura, *Appl. Polym. Symposium 28*, 713 (1976).
168. R. H. Marchessault and J. A. Howsmon, *Text. Res. J. 27*, 30 (1957).
169. G. Cuvelier and R. Aubry, *Chimie et Industrie 89*, 571 (1963).
170. R. Ferrus and P. Pages, *Cellulose Chem. and Techn. 11*, 633 (1977).
171. P. H. Hermans and A. Weidinger, *J. Appl. Phys. 19*, 491 (1948).
172. B. E. Dimick and R. H. Atalia, 169th National Meeting of the American Chemical Society, Philadelphia, 7-11 April 1975.
173. P. H. Hermans, in *Physics and Chemistry of Cellulose Fibers*, Elsevier, Amsterdam (1949).
174. J. W. S. Hearle and R. Greer, *Textile Progress 2*, 128 (1970).
175. J. H. Wakelin, H. S. Virgin, and E. Crystal, *J. Appl. Phys. 30*, 1654 (1959).
176. L. E. Alexander, *X-ray Diffraction Methods in Polymer Science* (L. E. Alexander, Ed.), Wiley, New York p. 180 (1969).
177. N. B. Patil, N. E. Dweltz, and T. Radhakrishnan, *Text. Res. J. 32*, 460 (1962).
178. M. Sotton, *Industrie Minérale, III 6*, 1 (1975).

179. L. M. Meneses Guimaraes De Almeida, Etude de l'influence des conditions de mercerisage sur les propriétés des fibres cellulosiques, Thesis, Mulhouse, France (1978).
180. P. K. Chidambareswaran, N. B. Patil, and V. Sundaram, *J. Appl. Polymer Sci. 20*, 2297 (1976).
181. S. H. Zeronian and K. E. Cabradilla, *J. Appl. Polym. Sci. 17*, 539 (1973).
182. H. J. Phillips, M. L. Nelson, and H. M. Ziffle, *Text. Res. J. 17*, 585 (1947).
183. R. S. Orr, A. W. Burgis, J. J. Creely, T. Mares, and J. N. Grant, *Text. Res. J. 29*, 355 (1959).
184. M. L. Nelson and V. W. Tripp, *J. Polym. Sci. 10*, 577 (1953).
185. O. A. Battista, S. Coppick, J. A. Howsmon, F. F. Morehead, and W. A. Sisson, *Ind. Eng. Chem. 48*, 333 (1956).
186. E. A. Immergut and B. G. Ranby, *Ind. Eng. Chem. 48*, 1183 (1956).
187. G. M. Venkatesh and N. E. Dweltz, *J. Appl. Polym. Sci. 20*, 273 (1976).
188. T. Radhakrishnan, B. V. Iyer, G. S. Viswanathan, and H. Wakeham, *Text. Res. J. 29*, 332 (1959).
189. J. M. Preston, *Trans. Faraday Soc. 29*, 65 (1933).
190. J. J. Hebert, L. L. Muller, R. J. Schmidt, and M. L. Rollins, *J. Appl. Polym. Sci. 17*, 585 (1973).
191. L. Fourt and A. M. Sookne, *Text. Res. J. 21*, 469 (1951); *ADR 43*, 304 (1954).
192. A. Anton, K. A. Johnson, and P. A. Jansson, *Text. Res. J. 48*, 247 (1978).
193. L. Fourt and H. J. Elliot, *Text. Res. J. 25*, 11 (1955).
194. O. Mecheels, *Melliand Textilber. 13*, 645 (1932).
195. D. Bechter, *Textil Praxis Int. 31*, 1431 (1976); *32*, 178 (1977).
196. J. J. De Boer, *Text. Res. J. 43*, 141 (1973).
197. L. Rebenfeld, *Text. Res. J. 28*, 462 (1958).
198. H. Wakeham and N. Spicer, *Text. Res. J. 21*, 187 (1951); *25*, 585 (1955).
199. A. W. MacDonald, R. S. Orr, G. C. Humphreys, and J. N. Grant, *Text. Res. J. 27*, 641 (1957).
200. R. S. Orr, A. W. Burgis, F. R. Andrews, and J. N. Grant, *Text. Res. J. 29*, 349 (1959).
201. S. H. Zeronian, K. W. Alger, and K. E. Cabradilla, *J. Appl. Polym. Sci. 20*, 1689 (1976).
202. J. J. De Boer and H. Borsten, *J. Text. Inst. 53*, T495 (1962).
203. R. Freytag, L. M. Meneses Guimaraes de Almeida, and J. J. Donzé, *Textilveredlung 13*, 486 (1978).
204. B. R. Shelat, T. Radhakrishnan, and B. V. Iyer, *Text. Res. J. 30*, 836 (1960).

205. R. S. Orr, A. W. Burgis, L. B. Deluca, and J. N. Grant, *Text. Res. J. 31*, 302 (1961).
206. W. A. Reeves, W. G. Sloan, and A. S. Cooper, *ADR 53*, 958 (1964).
207. P. W. Sherwood, *Textil Praxis 20*, 482 (1965).
208. R. L. Colbran and T. M. Thompson, *J. Text. Inst. 58*, 385 (1967).
209. C. F. Goldthwait, *Text. Res. J. 35*, 986 (1965).
210. D. A. Glibbens and B. P. Ridge, *J. Text. Inst. 18*, T135 (1927).
211. R. W. Jacoby, *ADR 35*, 56 (1946).
212. J. Boulton and T. H. Morton, *JSDC 56*, 145 (1940).
213. P. Kubelka and F. Munck, *Z. Techn. Physik 12*, 593 (1931).
214. C. F. Goldthwait, *Text. Res. J. 47*, 632 (1977).
215. G. Goldfinger, *Text. Res. J. 47*, 633 (1977).
216. D. Bechter, D. Fiebig, and S. A. Heap, *Textilveredlung 9*, 265 (1974).
217. K. Bredereck, *Melliand Textilber. 59*, 648 (1978).
218. K. Bredereck, *Textilveredlung 13*, 498 (1978).
219. H. Fischer, *Textilveredlung 13*, 507 (1978).
220. P. Grünig, *Textilveredlung 13*, 510 (1978).
221. M. Rösch and W. Langmann, *Melliand Textilber. 49*, 311, 450 (1968).
222. M. Rösch, *Melliand Textilber. 40*, 1306 (1959).
223. G. Dollinger, M. Mohaupt, and M. Rösch, *Melliand Textilber. 41*, 1398 (1960).
224. Hoechst A.G., British Patent 1,430,118 (31 March 1976).
225. V. A. Shenai, *Inter. Dyer 158*, 246 (1977).
226. C. Duckworth and I. Rusznák, *Textile Month 1*, 60 (1976).
227. C. Duckworth and L. M. Wrennall, *JSDC 93*, 407 (1977).
228. E. K. Boylston and J. J. Hebert, *Text. Res. J. 49*, 317 (1979).
229. C. Duckworth, British Patent 1,388,263 (26 March 1975).
230. Matther and Platt, British Patent 1,485,756 (14 September 1977); U.S. Patent 4,095,944 (20 June 1978).
231. Heberlein A.G., U.S. Patent 3,960,484 (1 June 1976).
232. D. Bechter, *Textil Praxis Int. 33*, 75, 177 (1978).
233. Brugman Machinefabrik, *Melliand Textilber. 60*, 734 (1979).
234. D. Bechter, *Textilveredlung 13*, 490 (1978).
235. D. Bechter and G. Kurz, *Textil Praxis Int. 34*, 965, 1369 (1979).
236. G. Golbs and K. Pankow, *Textiltechnik 28*, 369 (1978).
237. L. A. Gotovtseva, A. I. Bat'Kov and A. I. Maklashin, *Tekstil'naya Promyshlennost' 34*, 54 (1974).
238. E. Feess, *Textil Praxis 23*, 335, 387, 469 (1968).
239. H. Flecken, *Textil Praxis 25*, 305, 365, 488, 562 (1970).
240. L. Martin and G. Oehler, *Melliand Textilber. 53*, 1139 (1972).

241. G. Fisher, *J. AATCC 1*, 407 (1969).
242. R. H. Peters, *Textile Chemistry*, Vol. 2 (R. H. Peters, Ed.), Elsevier, Amsterdam, p. 330 (1967).
243. L. Burtal, *L'Industrie Textile 903*, 479 (1962).
244. E. Gassmann, *Textilveredlung 13*, 514 (1978).
245. J. Tomek, *Melliand Textilber. 54*, 637 (1973).
246. A. C. Tate, *Textile World 93*, 84, 95 (1943).
247. E. M. Leimbacher, *Inter. Dyer 154*, 375 (1975).
248. *Textilveredlung 13*, 527 (1978).
249. V. Biondic, *Wirk. u. Strick. Technik* pp. 6-7 (1975).
250. *Melliand Textilber. 55*, 58 (1974).
251. *Inter. Dyer 158*, 495 (1979).
252. H. Weber, *Textilveredlung 9*, 276 (1974).
253. E. Gassmann, *Textilveredlung 13*, 516 (1978).
254. *Textilveredlung 13*, 528 (1978).
255. P. F. Greenwood, *Text. Inst. and Ind. 14*, 373 (1976).
256. Lindauer Dornier GmbH, *L'Industrie Textile 1062*, 707 (1976).
257. E. Wirth, *Textilbetrieb 95*, 56 (1977).
258. H. Weber, *Textilveredlung 13*, 519 (1978).
259. *Melliand Textilber. 59*, 225 (1978).
260. G. Euscher and E. Wirth, *Textil Praxis Int. 35*, 194 (1980).
261. *Inter. Dyer 158*, 494 (1979).
262. *Melliand Textilber. 61*, 266 (1980).

4

LIQUID AMMONIA TREATMENT OF TEXTILES

CATHERINE V. STEVENS Sandoz Colors and Chemicals, East Hanover, New Jersey

LUIS G. ROLDAN(-GONZALEZ) J. P. Stevens & Co., Inc., Greenville, South Carolina

1. Introduction 168
2. Properties of Liquid Ammonia 168
3. Effect of Liquid Ammonia on Structure and Morphology of Cotton 170
4. Effect of NH_3 on Rayon or Mercerized Cellulose 175
5. Liquid NH_3 Treatment of Cellulosics as a Substitute for Mercerization or as a Pretreatment for Easy Care Finishing 176
   - 5.1 Treatment of yarn with liquid ammonia 177
   - 5.2 Treatment of fabrics with liquid ammonia 180
6. Liquid NH_3 in the Application of Topical Finishes 190
   - 6.1 Crease-resist finishing 190
   - 6.2 Flame-resist finishing 192
7. Dyeing from Liquid Ammonia 193
8. Liquid Ammonia Treatment of Noncellulosic Fibers 195
   - 8.1 Wool 195
   - 8.2 Nylons 198
9. Equipment 198
10. Future of Liquid Ammonia Processing 198

References 199

1. INTRODUCTION

Although basic research into the effect of liquid ammonia on cellulose was published as early as 1935 [1], it was not until the 1960s that the use of liquid ammonia as a wood plasticizer was disclosed [2] and that its commercial utility in the treatment of cotton-containing yarns and fabrics began to be realized [3]. By the mid-1970s liquid ammonia treatments had become accepted as a substitute for mercerization on cotton sewing threads [4] and as a means to improve the easy care properties of heavyweight cotton fabrics, such as denim [5].

2. PROPERTIES OF LIQUID AMMONIA

Liquid ammonia has received much attention as a nonaqueous reaction medium because of its moderate dielectric constant, high dipole moment, ability to hydrogen bond, and relatively high basicity [6]. Some of the physical constants of anhydrous ammonia are summarized in Table 4.1. The purity of commercial grades of anhydrous ammonia

Table 4.1 Physical Constants of Anhydrous Liquid Ammonia

Molecular weight	17.03
Boiling point at 1 atm	-33.35°C
Freezing point at 1 atm	-77.7°C
Latent Heat of Vaporization at 1 atm, -33.4°C	327 Cal/g
Specific Heat, Cp at 20°C	1.125 Cal/g/°C
Viscosity at -33.5°C	0.266 centipoise
Specific gravity at -40°C	0.690
Specific gravity at -33.35°C	0.674
Specific gravity at 0°C	0.639
Specific gravity at 40°C	0.580
Dielectric constant at -50°C	22.7
Dielectric constant at 5°C	18.9
Electrical conductivity	$1 \times 10^{11} \text{ohm}^{-1} \text{cm}^{-1}$
Vapor pressure of liquid, $\log_{10} P_{mm}$	$9.9508 - \frac{1473.17}{T - 3.8603} \times 10^{-3} T$

Source: Refs. 6, 7, 9.

generally excees 99.5%, with the major contaminants being water, oil, and noncondensable gases [7, 8].

Several factors, especially, must be considered with regard to safety in working with and handling liquid ammonia. In addition to the corrosive action of high concentrations of ammonia on human tissue, liquid ammonia can cause severe injury by freezing the tissue. Also, although ammonia does not represent a serious fire hazard, mixtures of 15-28% ammonia in air can be ignited by sparks or temperatures exceeding 1200°F [7]. Since explosive compounds can form when ammonia and mercury are combined, instruments exposed to ammonia should not contain mercury. Liquid ammonia is hygroscopic [6]; although anhydrous ammonia will not affect most common metals, moist ammonia vigorously attacks copper, silver, zinc, and their alloys. Thus, iron and steel are normally recommended as materials of construction for objects that come into contact with ammonia [8].

Lagowski and Moczygemba [6] have reviewed the solubility of various chemical types in liquid ammonia. Among the covalent substances,

> Aromatic and halogenated aliphatic hydrocarbons are generally more soluble than aliphatic hydrocarbons.
>
> Nitro substitution for hydrogen tends to increase the solubility of organic radicals.
>
> Molecules containing highly polarizable atoms (e.g., iodoform) or functional groups (e.g., carbonyl moiety) are expected to show increased solubility in ammonia.
>
> Groups capable of forming H bonds to ammonia (e.g., alcohols, primary and secondary amines) or of being H bonded by ammonia (e.g., ethers, tertiary amines, pyridine) exhibit enhanced solubility.
>
> Molecules possessing acidic hydrogen atoms (e.g., carboxylic acids, phenols, imides) react with ammonia to form ammonium salts that are soluble. However, the highly oxygenated compounds (e.g., di- or polyacids) are less soluble.

Of the electrolytes,

> Metal halides display the following order of increasing solubility: $F^- < Cl^- < Br^- < I^-$.
>
> Salts containing highly oxygenated anions (e.g., SO_4^{-2}, PO_4^{-3}, CO_3^{-2}, CrO_4^{-2}, AsO_4^-, BO_2^-, SO_3^{-2}, O_2^-, OH^-, and dicarboxylates) are virtually insoluble, except for the perchlorates and nitrates.
>
> Compounds containing cations with relatively high charge density (e.g., Li, Mg, Ca salts), and salts containing easily polarizable anions (e.g., NaI, RbI) dissolve to form ammoniates.
>
> Generally the NH_4 salt of a given anion is the most soluble salt in liquid ammonia.

3. EFFECT OF LIQUID AMMONIA ON STRUCTURE AND MORPHOLOGY OF COTTON

In its native state cellulose crystallizes in a lattice designated as cellulose I. Strong swelling agents are capable, however, of altering this crystal structure. For instance, cellulose II describes the lattice found in regenerated or alkali-mercerized cellulose. Treatment with liquid ammonia or hot glycerol can result in lattices known as cellulose III and IV, respectively.

Among the various forms of ammonia, the liquid appears to be unique in its swelling action on cellulose and its effect on the crystal structure. For instance, dry cotton has been reported to slowly take up 4% ammonia gas at room temperature and atmospheric pressure, with no change in the x-ray diffraction diagram. In concentrated aqueous ammonia, cellulose swells slightly, but again the crystal structure is not affected [10].

Anhydrous liquid ammonia penetrates cellulose very rapidly and complexes with the hydroxyl groups after breaking hydrogen bonds in the crystalline as well as in the less ordered regions of the fiber. The ammonia molecule is a relatively small molecule, not large enough to dissolve the cellulose, but capable of increasing distances between cellulose chains through its penetration into the crystallites.

Much of this can be deduced from the earliest published studies on the action of liquid ammonia on cellulose, among them that of Barry, Peterson, and King [11] in 1936. Working with ammonia condensed at the temperature of dry ice/acetone (approximately -50°C) and treatment times of 5-10 hr on ramie, they found an increase in the 101 plane spacing from 6.28 Å for the untreated cellulose to 8.88 Å in the ammonia-swollen cellulose. The ammonia cellulose complex contained approximately 1 mol of ammonia per anhydroglucose unit (AGU). Upon heating the complex at 105°C for several hours, it lost all its ammonia and underwent a structural change, the 101 spacing decreasing to 7.9 Å. Alkali-mercerized cellulose exhibited a 101 spacing of 7.7 Å. The calculated unit cell volumes were: native cellulose, 671 $Å^3$; NH_3 swollen, 801 $Å^3$; NH_3 treated and dried, 702 $Å^3$.

The authors suggested a summary of the transformations occurring in this system, which reads as follows when the currently accepted nomenclature for the various cellulose modifications is used:

$$\text{cellulose I} \underset{H_2O}{\overset{NH_3\ (\text{liq})}{\rightleftarrows}} NH_3\text{ - cellulose} \underset{NH_3\ (\text{liq})}{\overset{\Delta}{\rightleftarrows}} \text{cellulose III}$$

Thus, the ammonia cellulose complex could be attained either from cellulose I or III. Conversion of ammonia cellulose to cellulose III was obtained by heating or prolonged storage in paraffin oil; conversion to cellulose I resulted from immersion of the complex in water, dilute acetic acid, or dilute or concentrated ammonia. Cellulose III appeared to be a stable end product, since samples treated with either saturated aqueous ammonia or 1% acetic acid for several hours at room temperature, or heated at 105°C, exhibited no change in x-ray diffraction pattern. Although cellulose III appeared to give a striking resemblance to regenerated or alkali-mercerized cellulose (termed by Barry et al. as *hydrated*, now known as cellulose II), it was obviously different in that their cellulose III could be reconverted to cellulose I, while cellulose II could not be similarly reconverted.

Apparently working concurrently with Barry et al., Hess and Trogus [1] also found an x-ray diffraction pattern on ammonia-swollen cellulose similar to that found by the previous authors. In contrast, however, they found no appreciable swelling. Also, they were unable to obtain the original cellulose I structure by treatment with concentrated aqueous ammonia or with methanol.

Later, Hess and Gunderman [12] showed that the ammonia cellulose complex structure could vary depending on the temperature. Below -30°C, a so-called ammonia cellulose II was formed, and around or above -30°C the product was called ammonia cellulose I; ammonia cellulose II could be converted to the ammonia cellulose I complex by raising the temperature to around -30°C. Meyer [10] has suggested that the ammonia cellulose II complex consists of 2 mols of ammonia per AGU, while the analogous I complex possesses only 1 mol of ammonia.

Since the ammonia cellulose I complex as an intermediate for cellulose III is the most technologically interesting, Warwicker [13] has proposed a unit cell for this complex based on the work of Hess and Gunderman [12], assuming that the main equatorial reflections are the 101, $10\bar{1}$, and 002 reflections of a monoclinic cell. A comparison of this unit cell with that of cellulose I is shown in Fig. 4.1. It is

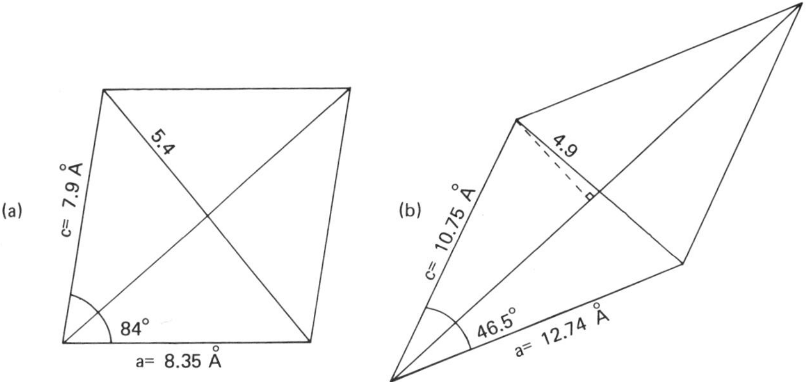

Figure 4.1 (a) Unit cell of cellulose I; (b) unit cell of ammonia cellulose I. (*Source:* Ref. 13.)

evident that the ammonia molecules must have penetrated and caused expansion between the (101) planes; very little penetration has occurred between the (10$\bar{1}$) planes. Warwicker [13] claimed that the swelling process resembles that taking place in alkali mercerization, the major difference being that the larger sodium hydroxide molecule causes greater intrafibrillar swelling than does liquid ammonia.

Clark and Parker [14], working with ammonia at -75°C for 1 hr, found a threefold increase in thickness of ramie fibers. The fact that the fibers reverted to normal size upon removal of much of the ammonia, although ammonia cellulose I was still present, has been taken as evidence that liquid ammonia causes interfibrillar swelling [13]. Clark and Parker found a larger 101 spacing (10.3 Å) than did Barry et al. for the ammonia-swollen cellulose, but upon allowing the ammonia to evaporate for 45 hr, they found a similar value for the cellulose III (101) spacing (7.7 Å). In contrast to the findings of both Barry et al. [11] and Hess and Trogus [1], complete reversion of the ammonia cellulose complex obtained from cellulose I was found only upon treatment with concentrated aqueous ammonia; incomplete reversion occurred using water or dilute acetic acid. No reversion was found if the starting material was mercerized cellulose (cellulose II).

Cellulose III was again obtained by slowly evaporating the ammonia from the swollen native or regenerated cellulose or by heating at 105°C. Treating the cellulose III from native cellulose with boiling water or concentrated aqueous ammonia resulted in only about 40% reversion to cellulose I, while little or no reversion took place in the case of cellulose III from mercerized cellulose. Sodium hydroxide of mercerizing strength converted cellulose III from both sources to cellulose II.

Much of the seemingly conflicting evidence presented by the earliest authors stems not only from problems of nomenclature and limitations of the characterization techniques at the time, but also possibly from failure to consider the effects of temperature and time on the complex formation. The role of small amounts of moisture absorbed by the fibers, moisture that is capable of plasticizing and causing recrystallization, was probably not well understood at that time; similarly, the speed of ammonia evaporation could have caused variations in the extent of the modifications obtained.

The rapidity of the action of liquid ammonia on cellulose has been demonstrated in several studies, e.g., that of Lewin and Roldan [15]. After only 2 sec immersion in ammonia at approximately -30°C, the beginning of the cellulose I-III transition could be seen; after 10 sec immersion, the cellulose III lattice already predominated, and after 5 min immersion a minimum of cellulose I was present. Pandey and Nair [16] found that the maximum effect of ammonia (at temperatures ranging from -20°C to -72°C) on fiber tenacity and other mechanical properties occurs within 10 min. Treatments at -20°C for 10 min produced greater improvement in fiber tenacity than those at -40°C or -72°C; however, moisture regain, accessibility, and infrared crystallinity index showed a greater degree of change in fine structure for samples treated at -72°C compared to -20°C. Electron microscope observations of the -40°C and -72°C samples confirmed the occurrence of both inter- and intracrystalline swelling, without the loss of fibrillar structure.

Jung and Benerito [17] claimed that the extent of conversion of cellulose I from cotton to cellulose III after 3-9 sec immersion in liquid ammonia was more complete the lower the temperature of a subsequent quenching; temperatures in this study ranged from ambient to -196°C. The explanation offered was that the more exothermic the heat changes at the cotton surface could be maintained, the more cellulose III could be formed.

Wellard [18] has proposed that although the equatorial reflections from cellulose III are consistent with a monoclinic, hexagonal, or tetragonal unit cell, consideration of the calculated unit cell volumes and comparison of these with the unit cell volumes known for cellulose I and II leads to a choice of the hexagonal cell as the correct one. The density obtained for cellulose III prepared from bacterial cellulose reportedly confirms this choice.

Lewin and Roldan [15] have presented a phase diagram for the various ammonia cellulose I transitions (Fig. 4.2). The four major phases obtainable from cellulose I are shown at the corners of a tetrahedron. In the base are cellulose I (I), cellulose III (III), and disordered cellulose (D); the fourth corner, lying above this base plane, represents the ammonia cellulose complex (A-C). It was stressed that both I and III represent phases of high crystalline order but contain minor fractions of disordered cellulose, while D represents 100% disordered cellulose. Cellulose III is obtained by evaporating with heat the ammonia from A-C, which can be formed from the liquid ammonia treatment of I, III, or D. A transition from D to III is possible on application of dry heat, and from D to I with moist heat, while the reverse transitions are impossible unless a strong swelling agent like ammonia is used. The transition from III to I is possible through the application of water and heat, or by prolonged soaking in water at ambient conditions. The various transitions are influenced by time, temperature, and relative humidity; thus intermediate stages can be obtained which are characterized not only by the relative amounts of the phases present in the specimens, but also by the degree of order achieved in the crystalline phases.

The same authors [15] deduced from x-ray diffraction meridional reflections that the length of the cellulose III crystallites is smaller than those of cellulose I (14-23 Å versus 45-74 Å). This means that in cellulose III a crystallite is composed of no more than 16 chains, 12 of which are accessible by water molecules surrounding the

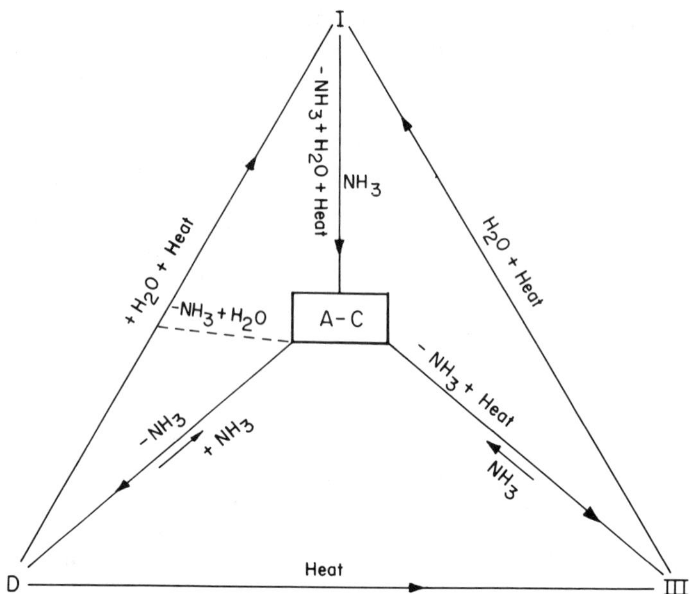

Figure 4.2 Phase diagram of ammonia cellulose (A-C), disordered cellulose (D), cellulose I (I), and cellulose III (III). A-C is the vertex of a tetrahedron and is placed above the plane of the paper.

elementary fibril. Thus water molecules play an important role in the III-I transition, due to enhanced mobility of the water molecules in the system.

A decrease in cotton fibril size upon liquid ammonia treatment has also been pointed out by Hebert et al. [19]. Measurements were made by electron microscopic observation of stained cross sections. A decrease in fibril size from 28-33 Å to 17-22 Å was found, even though the cellulose I structure was claimed to have been maintained.

Liquid ammonia treatment appears to result in, at most, small decreases in DP of cellulose. Pandey and Nair [16] found no change in DP of cotton after a 10 min treatment at -40°C, and a 20% drop after 2 hr treatment. Wadsworth et al. [20], working with wood pulp at about the same temperature for 1 hr, found less than a 10% decrease in DP.

Microscopic observations indicate [13] that convolutions normally present in cotton fiber are removed by liquid ammonia treatment if the ammonia is evaporated, but may be present if water is used to remove the ammonia. With regard to fiber cross sections, ammonia swelling in general results in thicker cell walls and more elliptical forms than the untreated fibers, which are predominantly flat or oblong in cross section [13, 21]. If water is used to remove the ammonia, the lumens are markedly more open than if the ammonia is simply evaporated. This is most likely due to the rapid gasification and expansion of ammonia when the fibers are placed in water.

Koura and Schleicher [22] have studied the effect of various ammonia solutions on the structure of cotton cellulose as inferred by changes in level-off DP, moisture retention, vapor sorption, and acetylation capability. It is claimed that with solutions of ammonia in water, DMSO, or formamide, structural changes can be detected with only 0.5 and 0.67 mols NH_3, respectively, per mol of solvent. With 55% ammonia-water solutions, it was claimed that structural changes equivalent to those obtained in anhydrous ammonia are seen. Treatments were carried out for 1 hr and were followed by washing in cold water. Alcohol-ammonia solutions caused no structural changes.

4. EFFECT OF NH_3 ON RAYON OR MERCERIZED CELLULOSE

Barry, Peterson, and King [11], as previously mentioned, found a strong similarity between cellulose II and III, both in x-ray deduced structure and in an increased chemical reactivity as represented by copper number; they recognized the differing nature of the two structures, however, in that cellulose III but not cellulose II could be reconverted to cellulose I. Also, while the same diffraction pattern was obtained on ammonia cellulose resulting from liquid ammonia treatment of ramie and alkali-mercerized cellulose, treatment of the latter with water yielded a structure which did not give an x-ray pattern sharp enough to measure. In a similar study, Clark and Parker [14], using concentrated aqueous ammonia rather than water for the reconversion, were able to obtain complete reversion of the ammonia-swollen native cellulose and only about 70% reversion of the ammonia-swollen mercerized cellulose. Sodium hydroxide of mercerizing strength converted cellulose III from either source (cellulose I or II) to mercerized cellulose (cellulose II).

Wellard [18] studied cellulose II and III structures obtained from a number of sources, including various rayon yarns as well as mercerized and native cellulose. He found small variations in the monoclinic angle for the various cellulose II samples, with β being greater for the mercerized specimens than for the regenerated ones. For the various cellulose III samples, no significant variation in unit cell parameters were found, although the intensity of the (020) reflection exhibited large variations, which indicated that the molecular arrangement in cellulose III is dependent upon the origin of the cellulose and its previous treatment. This was also taken as further evidence that the molecular arrangement in mercerized native cellulose and regenerated cellulose are, to some extent, different.

Using infrared spectroscopy of deuterated samples [23, 24], Marrinan and Mann confirmed that cellulose III prepared from cellulose I is structurally different from that prepared from cellulose II (viscose rayon). Supporting this was the behavior of the various forms toward water; cellulose III prepared from I reverted to cellulose I in a minimum of 10 hr, while cellulose III prepared from II reverted to cellulose II and was much more difficult to accomplish than the III-I reversion. The use of the subscript I or II to designate the origin of the cellulose was suggested.

The close resemblance of the IR spectra of unoriented cellulose III_{II} and oriented cellulose II obtained by Mann and Marrinan [23] indicated that the hydrogen bonding in the two forms is very similar. Intramolecular hydrogen bonds (i.e., in the 101 plane) were responsible for maintaining the found configurations.

The work of Roldan [25] supported the suggestion of Howsmon and Sisson [26] that cellulose IIIs are disordered forms of cellulose II. Thus, the form originally called cellulose III_{II} appears to be an interpenetration twinning of short-range order lattices of cellulose II. Several factors may contribute to this observation:

1. The inability of liquid ammonia to alter intramolecular hydrogen bonds in cellulose II, in contrast to its effect on cellulose I. This might be due to the fact that the cellulose II chain corresponds to an energy minimum, with stronger intramolecular hydrogen bonds than the other configurations.
2. The interference of the cellulose T_g on chain internal rotation at an application temperature which is at least 100°C lower than the assumed T_g.

Thus, Roldan concluded that the most obvious effect of liquid ammonia on cellulose II is its plasticizing activity. By its action as an intrafibrillar swelling agent, it is capable of releasing strains in the cellulosic lattice created during processing and can also increase the chemical accessibility of the cellulose II. This allows an improved alignment of the cellulosic chains and fibrils during the ammonia evaporation if stress is applied.

5. LIQUID NH_3 TREATMENT OF CELLULOSICS AS A SUBSTITUTE FOR MERCERIZATION OR AS A PRETREATMENT FOR EASY CARE FINISHING

The process of mercerizing cotton has been known for over 100 years [27], and involves the swelling of the fibers in a strong caustic solution (commercially, approximately 20% sodium hydroxide is used). It can be used to accomplish significant changes in the structure and morphology of cotton fibers, most notably the conversion of the native cellulosic crystalline lattice (cellulose I) to that of cellulose II. The degree of order of the fiber is, in general, decreased by mercerization treatments, thus increasing its accessibility to chemical reagents [28].

If the mercerization is carried out under tension, significant improvements in fiber tenacity and yarn luster can be accomplished. The increased tenacity is probably due to an increase in orientation of the fiber, while the increased luster is due to a combination of factors, including changes in cross-sectional shape and number of convolutions of the fibers, as well as to increased alignment and packing of the fibers in the yarn.

In practice, the improvements in strength are noticed mostly upon yarn treatments. Fabric mercerizing generally gives a nonuniform result, with its major effect on the surface of the fabric; this is acceptable since its primary aim is increased luster and dye yield [29]. Penetration of the liquor into the interior of the yarns and fabric is probably hindered by immediate swelling of the fiber surfaces, which causes a jamming effect; the application of tension to restrict shrinkage probably also restricts the amount of free space in the fabric available for swelling.

Thus, since the aims and results of yarn versus fabric mercerization are different, the treatment of the two substrates with liquid ammonia will be discussed separately. The similarity of mercerization and liquid ammonia treatment lies in the fact that both processes swell the fibers; the differences, however, between the two processes should be stressed.

5.1 Treatment of Yarn with Liquid Ammonia

The Prograde process for the liquid ammonia treatment of yarns was announced in 1969 by the Scottish firm of J. & P. Coats Ltd. [30, 31]. In this process yarn runs continuously through a liquid ammonia bath at -33°C (with approximately 0.7 sec contact time) into a hot-water bath (approximately 0.1 sec contact at 93°C) where it is stretched 10-30% as measured against its ammonia-swollen length. Subsequently the yarn is wound on a spool and dried continuously with hot air [30, 32]. It is claimed [32] that the Prograde process is one thousand times faster than traditional yarn mercerization processes and can be used to obtain the properties for which yarn mercerization is carried out, i.e., higher tenacity, luster, and dyestuff affinity. It is also claimed to be significantly more economical than traditional mercerization for a number of reasons, including the high process speed and a simplified package-to-package process sequence which leads to fewer yarn breaks and knots. Investment costs have been minimized by the elimination of costly ammonia recovery systems [32].

The Prograde process was originally developed primarily to replace mercerization in the preparation of sewing threads [30]. However, the luster, strength, dimensional stability, and uniformity imparted to cotton yarns are claimed to make the yarn very suitable for knitting applications also [3, 33].

In a later J. & P. Coats patent [34], it is claimed that pretreatment of cellulosics with moisture prior to liquid ammonia treatment increases the rate of the effect of ammonia on the material. In practice, this is claimed to be evidenced particularly by increased evenness of dye takeup. The amount of moisture should be greater than the normal moisture regain of the substrate, but not greater than 30% based on weight of dry material. Presumably, the water preswells the cellulose, increasing its accessibility in the short processing time in the anhydrous ammonia.

In an apparent attempt to enable increased flexibility of processing speeds, Omnium de Prospective Industriel (OPI) Cryochimie has proposed applying the liquid ammonia to the outside of a yarn bobbin (or fabric roll) and subsequently unwinding the package at a rate such that the unwound material is subjected to the treating fluid for the required minimum duration [35]. The treatment application of the liquid ammonia can be accomplished by immersing the bobbin or roll into a tank or by spraying it onto the outside of the bobbin or roll. Treatment times of at least 10 sec were claimed to offer maximum efficiency of treatment (as measured by barium hydroxide absorption).

A USSR procedure for liquid ammonia treatment of 18.5 and 50 tex open-end spun cotton yarns claims that processing under 8% tension for 1.5-2.5 sec with subsequent removal of ammonia by heat gives the best overall results [36].

Warwicker [13] has found that x-ray diagrams of fibers removed from Prograde treated yarns give a cellulose I diagram; if they have received the normal amount of tension, they also show an increase in orientation. He has also demonstrated that the increase in strength of fibers in Prograde treated yarns can be related to the x-ray angle ($\phi_{1/2}$), which is defined as the breadth of the azimuthal scan around the 002 arc on the x-ray diagram of the fiber at one-half maximum intensity. The reciprocal of this angle is referred to as the orientation parameter ($1/\phi_{1/2}$) and increases with increasing orientation. As the percent stretch applied to the yarn during processing increases, the orientation parameter also increases in a curvilinear fashion. As the orientation parameter increases, the extension of the yarn at break decreases while the breaking load increases. These results demonstrated the fundamental effect of the Prograde process on the fiber itself; since a reduced x-ray angle indicates a reduced spiral angle of the fibrils, it was claimed that the Prograde process results in an improved alignment of fibrils along the fiber axis, with a resultant decrease in potential points of failure.

Subsequent studies by U.S. Department of Agriculture workers confirm these observations [37, 38]. With both caustic mercerization and liquid ammonia treatment, increased tension applied to the swollen fibers decreased the x-ray angle. With both types of treatments, decreased orientation angle correlated with increased tenacity and initial modulus. However, since increased strengths were accompanied by losses in elongation to break, energy to rupture decreased as tension during the treatment increased. Caustic mercerization of fiber bundles produced smaller x-ray angles than did ammonia treatment, purportedly because sodium hydroxide is the stronger (albeit slower) swelling agent, and thus the swollen fibers could be stretched more at a given application of tension [39].

Studies of length changes accompanying liquid ammonia treatments and sodium hydroxide mercerizations of low-twist yarns under various loadings have given additional evidence of basic differences in mechanism of the two swelling agents [37, 38]. In both reagents, yarns and fibers contract when swollen under low loads and elongate when swollen under high loads; at all loads they elongate when the swelling agent is removed. In liquid ammonia, significantly higher (2-4 times) loads are required than in NaOH to reach the point where shrinkage changes to elongation. Under equal loads (greater than the minimum required to produce elongation when swollen), yarns swollen in liquid ammonia are always shorter than those swollen in caustic. Liquid ammonia treated yarns, however, elongate more than do the caustic-treated yarns when the swelling solution is removed, and thus length differences seen in the swollen yarns are not always apparent in the washed yarn.

Loading before swelling causes yarns to elongate slightly more than does loading after swelling has occurred [37, 38]; while on caustic-mercerized yarn this difference can persist into the final washed yarns, on ammonia treated yarn the elongation occurring upon ammonia removal essentially erases pre- and postloading differences. Yarns of immature cotton exhibit greater length changes than those of mature cotton at equal loads at every treatment stage.

Rayon fiber swollen in liquid ammonia at a low load [38] exhibits even greater length changes than does cotton fiber, indicating that length changes observed in swelling are a basic property of the cellulose polymer and are not solely dependent upon the gross spiral yarn or fiber geometry.

Moisture regain increases after both swelling treatments, but somewhat more for the caustic mercerized than for liquid ammonia treated yarns [38-40]. In both cases, regains reach a higher value if the swelling occurs before tension is applied, presumably because more swelling is allowed to occur under such conditions. Extent of acetylation was increased slightly more by caustic mercerization than by ammonia treatments [40]; slack-mercerized samples produced higher acetyl contents than did those mercerized under tension. However, iodine sorption value, claimed to be a sensitive measure of extent of mercerization, showed the largest increase for those samples treated in liquid ammonia with ammonia removal by evaporation, followed by caustic-mercerized samples, and last by those samples ammonia treated and water quenched [39, 40]. This is in agreement with the results of Schwertassek and Hochman [41], who found slightly higher iodine sorption values for commercially caustic-mercerized yarn than for Prograde treated yarn (70-80 for caustic mercerized yarn versus 67 for NH_3 treated yarn, untreated equaling 38). Generally, only minor differences in iodine sorption value were found between samples mercerized slack or under tension. Accessibility to deuterium exchange, acid hydrolysis, and reaction with formaldehyde appeared to be relatively insensitive to detecting variations caused by swelling conditions [39, 40].

Bredereck [42] has carried out investigations to compare the luster imparted by caustic mercerization versus ammonia treatment (laboratory and Prograde processing). Although both types of treatments enhance luster, caustic appeared to give the superior effect. The effect of liquid ammonia treatments on luster was dependent on the method of ammonia removal, with hot-water immersion superior to evaporation and steaming. Other workers [41] saw less difference in luster between caustic mercerized and Prograde treated yarns; microscopic examination of yarn cross sections revealed that while in the mercerized yarn 63% of the fibers had assumed a round cross section, only 26% of the fibers in the Prograde yarn exhibited this characteristic. Electron microscopy, however, revealed that the fibers in caustic-mercerized yarns often showed "scarred, spiral, longitudinal stripes"; such markings were found infrequently in Prograde yarns. Thus it was concluded that the enhanced luster of Prograde yarns is due at least in part to favorable, smooth fiber surfaces, although fewer fibers are converted to cylindrical forms.

A study has been reported [43] on the piece dyeing of fabrics knitted from caustic mercerized versus ammonia treated yarns. Caustic mercerization had a significantly greater effect on increasing color values than did liquid ammonia.

5.2 Treatment of Fabrics with Liquid Ammonia

Paralleling the development of the Prograde process for liquid ammonia treatment of yarns was the development of a process for liquid ammonia treatment of fabrics jointly by the Norwegian Textile Research Institute and the Norwegian Central Institute for Industrial Research [44]. This work was begun in 1963 and the necessary machinery was completed in mid-1970. World-wide licensing of this Tedeco process on woven cellulosic fabrics was taken over by the Sanforized Co. The process is now known as the Duralized or Sanfor-Set process, depending on the degree of stabilization achieved [5].

Impregnation of the dried and cooled fabric occurs in a bath of liquid ammonia, with squeezing to about 100% wet pickup. The fabric is exposed to the liquid ammonia for about 10 sec, after which it is essentially dried by being run over blanketed, steam-heated dry cans. The treatment chamber is kept under a slight vacuum to prevent escape of gaseous ammonia. The small amount (5-10%) of residual chemically bound ammonia in the fabric after drying is removed by a light steaming. About 90% of the total ammonia is recovered for reuse, with the remainder trapped in water for later sale as an industrial chemical or fertilizer [5, 44, 45]. The extent of fabric shrinkage is controlled by the time of exposure to the liquid ammonia [46]; about 80% of the fabric shrinkage occurs during the first 5-6 sec [45]. Warp shrinkage can additionally be controlled by fabric tension during processing [47] and also by compressive shrinkage subsequent to liquid ammonia treatment, a feature of the Sanfor-Set process.

Although these processes were originally intended as replacements for mercerization, the impetus for Sanfor-Set commercialization in the United States (in 1974) was the treatment of heavy weight cotton fabrics, e.g., denim and corduroy. This combination of liquid ammonia treatment and Sanforizing is claimed to give enhanced dimensional stability, softness, and smooth-drying properties, with reduced seam puckering, edge fraying, and leg twist in garments [48]. In 1977 it was reported that six complete Sanfor-Set lines were in operation in the United States, and an additional liquid ammonia range was operating in Norway [49].

In a modification of the treatment procedure for heavy fabrics, Burlington Industries has suggested the application of the ammonia to the back of the fabric with some type of metering roll, followed by drying and steaming [50]. This conserves the energy required to remove ammonia, and on pile fabrics prevents distortion of the face. Another suggestion to avoid pile distortion has been to stabilize the pile structure through a controlled resin treatment prior to ammonia treatment [51].

Thus it can be seen that, commercially, two alternative technologies exist for liquid ammonia treatment: one is a water-based process for removal of the ammonia from the swollen substrate; the other is a dry/steam process where ammonia is removed largely by dry heat, with the last traces removed by steam. At present, the water-based system is commercially feasible only for yarn treatment, as discussed previously. The Duralized and Sanfor-Set processes represent the second approach.

In principle, the last traces of ammonia could also be removed from the dried substrate by water rather than steam. In practice, however, this approach is not attempted [52].

Ironically, the bulk of the experimental evidence appears to indicate that fabric treated through the water-based approach would be much more similar to conventional caustic-mercerized goods than is fabric processed via the commercially available dry/steam process. For this reason, Heap urges that the term *mercerizing* be restricted to processes employing caustic soda or caustic potash as the swelling agent [52], and that convention is being followed here. Bredereck [53] has prepared an excellent summary comparing the effects of various types of mercerization and liquid ammonia treatments, a translation of which is given in Table 4.2. Certain points should be noted in amplification of this summary table.

1. Bredereck and Heap [54] have shown that the stress generated by the interaction of cotton with liquid ammonia was up to four times higher than with caustic soda, even though the shrinkage was comparable (Table 4.3). It is hypothesized that although NH_3 and NaOH both react with cellulose on an intramolecular basis, NH_3 molecules produce temporary cross-links between cellulose chains which inhibit stress relaxation in the fiber. The low temperature of the treatment may also contribute to this. Table 4.3 also demonstrates the rapid reaction rate of ammonia and the relative insensitivity of the ammonia treatment to the preparation state of the fabric.
2. All the mercerization and liquid ammonia treatments discussed apparently eliminate the rope marks which are often characteristic of rope-bleached cotton fabrics [52, 55].
3. Varying reports regarding the depth of shade of ammonia dry/steam treated fabrics have been seen, ranging from *as good as* mercerized to *worse than* untreated [52]. This variation is probably due to the critical effect of the amount of ammonia remaining on the fabric at the time it enters the steamer. Various workers [52, 56] generally agree that if the amount of ammonia is greater than 35%, the results will be essentially the same as the water-based ammonia treatment; if the amount is below 5%, the result is typical of the dry/steam process, i.e., essentially no improvement in dye pickup. Results may vary depending on the size of the entering molecule; Heap [52] claims that the dry/steam process appears to make the fiber more accessible to small molecules, less accessible to large ones.
4. Bredereck [57] reports that depth of shade at a given dyestuff concentration in the fiber can be increased by swelling treatments if the irregularity of the fiber is reduced, and especially if the internal light scattering is lowered, e.g., by closing the lumen; these both occur in conventional mercerization treatments.
5. Stabilization of the fiber structure by conversion of the ammonia-swollen fiber from cellulose III to I (accompanied by crystallization and relaxation) while the ammonia is being removed appears to impart dimensional stability to the substrate [58]. Thus the steaming step during the Sanfor-Set process

is extremely important, as dry evaporation alone of the ammonia does not stabilize the structure sufficiently (Fig. 4.3). Ammonia removal directly with hot water (as in the Prograde process for yarns) seems to be even more effective for imparting dimensional stability, comparable to hot mercerization. Similar stabilization can be obtained by carrying out the swelling treatment slack and then restretching. However, it is extremely critical that the restretching occur while the swelling agent is still present; the stabilization cannot be permanent if the shrunken structure has already begun to stabilize.

The NH_3 dry/steam treatment differs from the other swelling treatments discussed in that friction-related retentive forces are reduced since structure stabilization occurs in an essentially deswelled fiber [59], thus leading to improved flexural and crease recovery properties (Table 4.4). Barkhuysen [60] has also reported that a heat removal of ammonia from swollen fabrics produces a higher crease recovery angle (CRA) than does removal by water. Calamari et al. [55] also noted maximum improvements in CRA, flex life, and stiffness by simple evaporation of ammonia from the complex, rather than by water quenching.

The theoretical basis for the Duralized (NH_3 dry/steam) process on heavy cottons can thus be better understood, i.e., improved dry crease recovery properties without the need for a strength-impairing resin treatment. On lighter weight fabrics, a higher crease recovery level is required for easy care properties; it has been claimed that since the base dry CRA is higher on ammonia pretreated fabrics, less cross-linking is required, leading to greater retention of strength and abrasion resistance compared to durable press finishing of untreated goods [59].

Later work on commercial liquid ammonia or mercerizing equipment has failed to support this claim when actual durable press smoothness ratings (rather than CRA) are used as the basis of the evaluation. Jones et al. [61] have demonstrated that on bottomweight fabrics, Sanfor-Set pretreatment results in improved durable press ratings and dry CRA before and after subsequent finishing with a cross-linking resin. However, on lightweight sheeting and shirting fabrics, neither liquid ammonia pretreatment nor mercerization permitted the use of less cross-linking agent to achieve commercially acceptable durable press performance. At levels of cross-linking agent in the bath below 9% solids, both mercerization and liquid ammonia treatments tended to depress durable press ratings; at higher cross-linker levels, both pretreatments boosted durable press performance (Fig. 4.4a). While liquid ammonia pretreatment boosted dry CRA at all resin levels, it depressed wet CRA, which might explain the effect of the pretreatment on durable press performance. Mercerization was less effective

Table 4.2 Comparison of Various Swelling Processes

	Mercerization		Liquid ammonia methods	
	Conventional	Hot	NH_3/water	NH_3/dry/steam
Characteristics of swelling process				
Speed	Relatively slow	Relatively fast	Very fast, even on greige goods	
Degree	High	Decreases with increasing caustic temp.	Somewhat less than in hot caustic	
Evenness	Tight constructions may lead to uneven swelling	Good	Good	Good
Shrinkage forces	Relatively small shrinkage forces; textiles exhibit good extensibility in swollen condition		High shrinkage forces; may lead to difficulties in maintaining exact dimensions	
Properties of cotton textiles after the treatment				
Luster	large increase	perhaps somewhat better than conv. mercerization	improvement not quite as high as by merc.	only slight increase
Dye takeup	strongly increased	not quite as high as conv. mercerization	80-90% compared to mercerized material	slight or no improvement

Dye savings	Approx. 30-50%	Approx. 30-50%	similar or perhaps slightly less than merc.	somewhat worse than mercerization
Strength	Improvements in the treatment of yarn or knit goods, none in the treatment of fabric. Similar effect for all methods.			
Dimensional stability	Similar effect by all methods. On heavy and tight fabrics, ammonia treatments have the advantage.			
Resistance to deformation	Relatively stiffer and harsher in hand than untreated	Somewhat softer hand than by conv. mer.	Similar to hot mercerization	Softer and more resilient hand due to disintegration of frictional restraints
	Dry crease recovery angle (CRA) of textile is scarcely altered.			Distinctly increased dry CRA
Response to subsequent cross-linking	Less loss in tensile strength on NH_3-treated yarns or knitwear; only slight effect on NH_3-treated fabric.			Less loss in tensile and tear strength, abrasion resistance on fabrics, esp. in combination with minimal add-on techniques

Source: Ref. 53.

Table 4.3 Shrinkage Force Behavior of Cotton Fabric Strips in Liquid Ammonia (-40°C) and in 20% NaOH (20°C)

Fabric	Swelling liquor	Equilibrium shrinkage force (kp/cm^2)		Time (sec) to obtain shrinkage force yield of	
		Warp	Fill	50%	90%
Cotton printcloth					
Gray	NH_3	1.3	0.94	7.5	42
Gray	NaOH	0.369	-	324	660
Bleached	NH_3	2.0	0.45	2	24
Bleached	NaOH	0.590	-	15	216
Cotton poplin					
Gray	NH_3	2.15	1.48	5	71
Gray	NaOH	0.504	-	264	1080
Bleached	NH_3	3.25	0.58	2	62
Bleached	NaOH	0.890	-	9	198

Source: Ref. 54.

in boosting dry CRA, but depressed wet CRA similarly. In this work, both liquid ammonia and mercerization pretreatments resulted in significant strength retentions on resin finished goods, with no real differences between the two pretreatments. Liquid ammonia pretreatment appears to have a substantial advantage in improved abrasion resistance, measured by Stoll flex, accelerotor weight loss, and multiple laundering tests (Fig. 4.4b). It should be noted that strength retention resulting from commercial liquid ammonia treatment tends to be greater in the warp than in the filling direction due to the greater tension applied in the warp direction; the microstretch procedure referred to in the figures applies filling tension during preparation and/or finishing to impart extra strength in that direction.

Calamari et al. [55] reported that on both 100% cotton and polyester-cotton blends, destruction of the ammonia cellulose complex through evaporation or heating resulted in fabrics with substantially better physical properties after subsequent cross-linking than similar fabrics in which the complex was broken by water quenching.

Although they implied that this was due to maintenance of the cellulose III structure, the immersion in the aqueous cross-linking reagent probably forced additional reversion to cellulose I, and thus their effects were very similar to those expected from a process such as Duralized.

The beneficial effect of liquid ammonia pretreatment compared to mercerization on the abrasion properties of cross-linked cottons is consistently mentioned in studies on the effect of such pretreatments on a variety of substrates with several cross-linking systems, especially if the levels of cross-linking agent applied are not too high [52, 55, 61-66]. The reported effects on tear and tensile strengths appear to be more variable and less dramatic than the effect on abrasion properties and probably depend highly on the exact systems involved (i.e., the specific tensions, cross-linking system, substrate, etc.). Durable press improvements ranging from 0-1 rating unit as a result of liquid NH_3 pretreatment of cotton fabrics subsequently treated with a given level of cross-linking resin have been claimed [55, 61, 63, 64]; again, any improvement is probably very dependent on exact processing conditions, especially the cross-linking system. In general, the effect of liquid ammonia pretreatment on crease recovery appears to be to increase the ratio of dry/wet CRA [61, 64], at least with acid catalyzed cross-linking systems.

All the previous studies have pertained to treatment of cotton-containing textiles. It should be noted that a reference does appear to a USSR study on the effect of liquid ammonia treatment of viscose rayon staple fabric [67]. Significant increases in dimensional stability

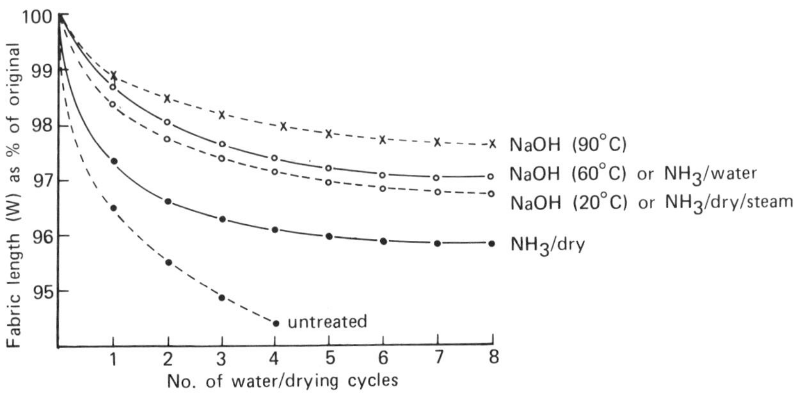

Figure 4.3 Shrinkage of cotton poplin as a result of water treatment and drying after constant length setting in 20% NaOH (1 min) or in liquid NH_3. (*Source:* Ref. 58.)

Table 4.4 Influence of Swelling Treatments on Bending Properties and Crease Recovery of Cotton Printcloth

	Cantilever bending stiffness ($cN \cdot cm \cdot 10^2$)				Crease recovery angle (degrees, W + F)			
	With tension		Slack		With tension		Slack	
Treatment	Warp	Fill	Warp	Fill	Dry	Wet	Dry	Wet
None	9.19	2.62	9.19	2.62	172	133	172	133
NaOH at 20°C	37.10	8.14	16.42	12.37	181	162	182	163
NaOH at 60°C	15.42	3.14	8.39	7.69	175	156	167	158
NaOH at 90°C	16.52	4.25	8.24	6.44	187	161	178	161
NH_3/water	27.46	5.88	10.39	5.75	195	169	186	168
NH_3/dry/water	10.88	3.33	4.46	2.98	210	167	220	169
NH_3/dry/steam	7.44	1.15	3.81	2.59	223	165	223	160

Source: Ref. 59.

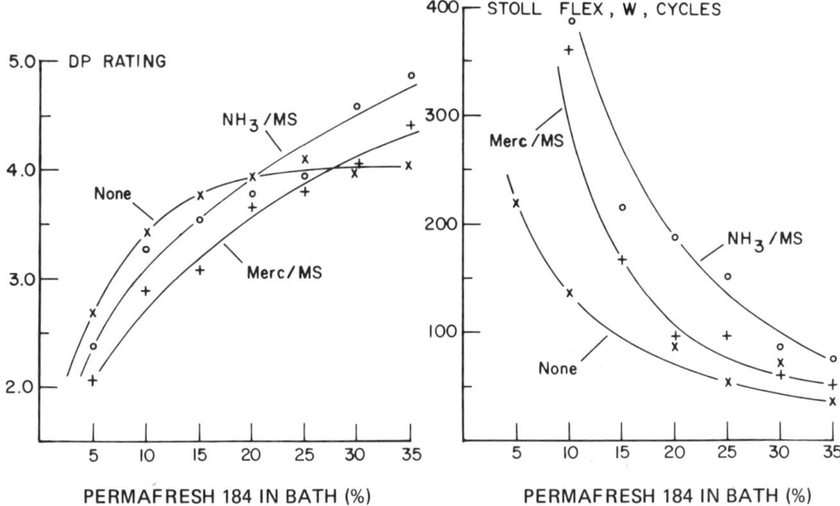

Figure 4.4 Dependence of various fabric performance properties on concentration of finish applied to cotton percale pretreated in three different ways: none, mercerized and microstretched, and liquid ammonia treated and microstretched. (a) Durable press performance (tumble drying); (b) flex abrasion (warp). (*Source:* Ref. 61.)

and dyeability with vat dyes were observed. Norwegian workers [68] have looked at the liquid ammonia treatment of rayon-cellulose wet-laid nonwoven webs. Optimum improvements in tear and tensile strength were found at around 70-75% ammonia with a furnish containing 60-70% rayon fibers. Although bulk of the webs was decreased, stiffness increased; aftertreatment in water led to a softer sheet, as well as to a reduced tensile energy absorption. The effect of liquid ammonia is probably due to two factors: (1) swelling of the rayon, leading to better bonding between fibers, and (2) shrinkage of the fibers in ammonia, leading to increase extensibility and tear strength.

Lambrinou [69] studied the liquid ammonia treatment of fabrics made from flax and tow. Degree of polymerization was reduced somewhat as a result of liquid ammonia treatment of both substrates, more so, however, on the flax. This reduction in DP was similarly reflected in tensile properties of the original fabric. After a few washes, however, the tensile strength of the ammonia treated substrate was higher than that of the original fabric, probably due to a higher shrinkage value for the ammonia treated goods, which also can lead to increased wet abrasion resistance. Dry crease recovery angles of the ammonia treated goods originally were almost doubled over those of the untreated substrates; after launderings the crease recovery values of the ammonia treated goods decreased about 20%, but were still significantly higher than those of the original goods.

6. LIQUID NH_3 IN THE APPLICATION OF TOPICAL FINISHES

6.1 Crease-Resist Finishing

The use of liquid ammonia as a solvent for the application of cross-linking agents to cotton-containing fabrics has been investigated by several groups. Bredereck and Pfundtner [70] cited nine N-methylol- or N-methoxymethyl-based cross-linkers which are soluble (to at least 0.5 M concentration) and stable in liquid ammonia solutions. These include the most commonly used cross-linkers, e.g., N,N'-dimethylol-4,5-dihydroxyethyleneurea (DMDHEU), N,N'-dimethylolethyleneurea (DMEU), hexamethoxymethylmelamine, and hexamethylolmelamine. They claim that thin layer chromatography (TLC) of the pure N-methoxymethyl or N-methylol compounds in this study, after having been dissolved in liquid ammonia for 4 hr, showed no evidence of amino methyl or amide groups, or of hexamethylenetriamine. Thus, no ammonolysis or dissociation of N-methylol groups appeared to have occurred. However, when commercial 50% aqueous solutions of DMEU or DMDHEU were dissolved in liquid ammonia and subsequently evaporated at room temperature, TLC of the residue showed the presence of NH-containing compounds, as well as hexamethylenetriamine. It was also claimed that an ammonolysis appeared to have occurred not in the solution of liquid ammonia but in the water-ammonia residue formed at room temperature during the evaporation process. If the liquid ammonia solution was kept at a lower temperature, TLC showed no evidence of any of the above reactions. N-methylol and free-formaldehyde determination of the commercial reactants after having been dissolved in liquid ammonia for up to 48 hr were also carried out; these showed unchanged N-methylol content but decreased free formaldehyde. Thus it was concluded that liquid ammonia reacts not with N-methylol groups but with free formaldehyde present in these reactants.

A variety of catalysts exhibit solubility in liquid ammonia. Of the conventional crease-resist-finishing catalysts, these include zinc nitrate, magnesium nitrate, ammonium nitrate, and ammonium chloride.

Bredereck and Pfundtner [70] saw essentially no improvement in CRA and relatively low reaction yields on fabrics treated with pure DMEU or commercial DMDHEU from liquid NH_3 if the NH_3 was simply evaporated at room temperature and the samples subsequently cured in dry heat (ammonium chloride or zinc nitrate as catalyst). Substantial increases in CRA and yield resulted if approximately 20% water was added to the NH_3 solution of the cross-linking agent or if the samples were steamed prior to heat curing so that a minimum amount of water (at least 15-20%) was condensed onto the fabric (Table 4.5).

Table 4.5 Finishing of Cotton Printcloth with DMDHEU from Liquid Ammonia: Influence of Steaming on Fabric Properties

Treatment	Water in fabric after steaming (%)	Crease recovery angle				Tenacity (g)		Extension (%)		%N
		Dry		Wet						
		Warp	Fill	Warp	Fill	Warp	Fill	Warp	Fill	
Untreated	-	42	44	74	85	243	251	15	20	-
Condensation with no presteaming	-	67	64	110	109	204	189	17	28	0.7
Steam at 98°C										
1 min	5.8	96	84	113	117	164	152	17	32	1.3
4 min	5.2	59	58	115	120	207	177	24	36	1.0
Steam at 60°C										
0.5 min	14.5	113	123	115	110	168	147	13	26	1.5
1 min	18	130	131	129	127	139	134	13	28	1.5
2 min	35	143	137	133	136	155	144	14	25	1.7
4 min	73	141	145	129	132	154	141	13	24	1.4

Bath: 100 g/l commercial DMDHEU + 10 g/l $Zn(NO_3)_2$ in liquid NH_3
Impregnation: 1 min, slack
Ammonia removal: 6 min at RT, stretch to within 5-8% of original dimensions. Dry 3 min at 70°C. Steam as indicated. Cure 5 min at 150°C.
Source: Ref. 70.

Steaming at 60°C resulted in more water in the fabric, and thus greater increases in CRA. These workers believe that under essentially anhydrous conditions, no diffusion occurs of the cross-linker into the fiber; cross-linking agents are predominantly deposited on the surface of the fibers. Short steaming procedures allow diffusion of the cross-linker into the interior of the fiber.

Lewin et al. [66] also observed that resin treatment of cotton from anhydrous ammonia led to low yields of resin insolubilization. Their explanation for this, however, was that an amination of the methylolated cross-linking agent had occurred; this was reported to have been indicated by an IR spectrum of trimethylolmelamine exposed to anhydrous ammonia. Although they did not report attempting a steaming step prior to heat curing, they did claim low yields even if the fabrics were resin treated by conventional aqueous systems, dried, and exposed to anhydrous ammonia before curing. Under such a set of conditions, diffusion of the cross-linker into the fibers should have occurred.

Bredereck and Pfundtner [70] claim that CRA and strength are somewhat better if NH_3 is evaporated in air before the steaming step, rather than if steam is used to remove the ammonia. This result correlates in general with the claims made by some workers for liquid ammonia treatment prior to DP finishing, i.e., the NH_3 dry/steam process versus the NH_3 water process [53, 55].

At low levels of DMDHEU treatment (measured by %N), application from water led to higher dry CRA than did application from anhydrous ammonia [70]. Wet CRA was essentially the same for all treatment routes. At a given dry CRA, DMDHEU-$Zn(NO_3)_2$ applied from water gave lower strengths than did application from liquid ammonia. This was probably an effect of the catalyst. In liquid NH_3, a $(Zn(NH_3)_6)(NO_3)_2$ adduct was isolated, which decomposed upon exposure to water, presumably to $Zn(OH)_2$, NH_4^+, and NO_3^-. Thus in ammonia processing, NH_4NO_3, not $Zn(NO_3)_2$, is probably the catalyst, possibly explaining the lower strength losses. In fact, an NH_4Cl-DMDHEU-water system gave a similar strength/CRA relationship to the $Zn(NO_3)_2$-DMDHEU-water systems. As was the effect for liquid NH_3 treatments prior to cross-linking, abrasion resistance was much better at equal CRA for the ammonia application systems compared to the aqueous application system.

6.2 Flame-Resist Finishing

The THPOH-NH_3 system has generally been regarded as a very suitable method for the flame retardant treatment of cellulosic fabrics. In a typical application of this system, the fabric is impregnated with a solution containing mainly a tetrakis(hydroxymethyl)phosphonium salt (e.g., the chloride or sulfate) which has been neutralized with NaOH or Na_2SO_3, and then the fabric is partially dried, passed through a chamber containing gaseous ammonia, washed, and oxidized with H_2O_2. It has been demonstrated by Calamari et al. that liquid ammonia can be substituted for the gaseous ammonia cure in such a system with no sacrifice in durability of the system [71]. The extremely rapid penetration of the liquid NH_3 solution offers the advantage of being able to increase processing speeds. Other claims that have been made for such a liquid ammonia substitution are that it permits greater latitude in drying conditions prior to ammonia curing and also that tensile strengths are improved [72].

Lewin et al. [66] attempted to apply from liquid NH_3 solutions flame-retardant finishes based on the in situ condensation of N,N',N"-trimethylphosphoric triamide (or its methylolated derivative) with trimethylol melamine. As in their work with durable press resins, reaction yields and, thus flame resistance, were unsatisfactorily low whenever the flame retardants were applied from liquid ammonia, even if a steaming step was included in the processing sequence. Again the effect was attributed to formation of less reactive, aminated products in the liquid ammonia.

Anhydrous ammonia aftertreatment of flame-retardant cotton and cotton blends did not impair the flame-retarding performance, but its effects on physical properties of the treated fabrics, including stiffness, was minimal [66]. Pretreatment of the fabrics prior to the application of the flame-retarding chemicals remained the most suitable approach to improving the properties of the flame-retardant cotton fabrics in their study. It was claimed that by immersing the fabrics in liquid ammonia and exchanging the ammonia with water prior to the application of the flame-retardant chemicals, stiffening of the fabric and losses in tear strength and flex abrasion were significantly reduced without impairing the flame-resistant performance of the fabrics.

7. DYEING FROM LIQUID AMMONIA

Because of its swelling ability, its speed of penetration into a variety of fibers, and its potentially unique solvent properties, a good deal of work has been carried out to determine how liquid ammonia can be used to advantage in a wide range of dyeing applications. For instance, Barkhuysen [73-76] investigated the dyeing of cotton with reactive dyes from liquid ammonia. None of the various types of reactive dyes examined produced satisfactory washfastness properties without the application simultaneously or subsequently of a resin finish. Simultaneous application of 10% DMDHEU resin, a catalyst (preferably NH_4Cl or NH_4NO_3), and the reactive dye from liquid ammonia for 60 sec was found to result in acceptable CRA, bursting strength, and depth of color. Adding water to the soaping formulation improved CRAs and fabric hand, but the effect on K/S values varied, decreasing with as little as 5% addition of water, then increasing [74]. A small amount of aqueous NaOH added to the liquid ammonia dyebath resulted in approximately a 10-fold increase in K/S values for the dyed cotton fabrics and an average sixfold increase in the amount of dye on the fiber compared to fabrics dyed the same way but with no NaOH in the dye bath [75].

The same investigator also showed that wool could be dyed with reactive dyes from liquid ammonia in short reaction times with high degrees of fixation [76]. Levelness was improved by the addition of 10% water to the liquid ammonia. Removal of ammonia from the fabric was accomplished by steaming.

Wade [77] has reported some work in which as much as 15-20% chloroform could be trapped in kiered yarn by swelling the yarn in a liquid ammonia-chloroform solution, then allowing the ammonia to evaporate. To obtain the maximum amount of inclusion, the conversion from cellulose I to III should be as complete as possible and the included chemical that replaces the liquid ammonia should not cause premature deswelling. This same technique was later utilized to dye cotton and cross-linked cotton with a variety of dyes, some of which are not traditionally used on cotton, e.g., acids, pigments, and basics [78]. To be useful in such a process, the dyes must be stable and soluble in liquid ammonia, as well as exhibit a partition coefficient which would enable the dyestuff to prefer the substrate. It was noted that regular vat dyes are not soluble in liquid ammonia, but vat esters are soluble and can be utilized in the process.

Dyeing from liquid ammonia is the subject of at least one patent which claims the utility of some unique combinations of dyestuff and fabric [79]. For example, triacetate was dyed deeply with direct and reactive dyes in liquid ammonia solution, but not in conventional aqueous baths. Depth of shade increased with increasing concentration of dyestuff in the liquid ammonia and with increasing time of immersion in the dyebath only up to 20 sec. No difference was seen in the depth of shade achieved on mercerized versus unmercerized cotton using liquid ammonia dyebaths.

A commercial system for dyeing from liquid ammonia has been promoted by Kane Industries [80]. Known as the rapid anhydrous method (RAM) dyeing system, it involves padding fabric, yarn or tow with the dye dissolved in liquid ammonia at -40°C at speeds of 80-120 yd/min, steaming the substrate for 3-30 sec to remove the ammonia and set the color, and finally afterwashing to remove unfixed dyestuff. It is claimed to be compatible with 95% of modern conventional dyestuffs and can be used on all dyeable fibers except wool and acetate. Advantages of the system are claimed to be primarily its continuous and rapid nature, which is especially interesting for knit fabrics, and furthermore its possibilities for savings in dye-assist chemicals.

OPI Cryochimie has been issued several patents for dyeing processes in which the substrate (natural or synthetic, but particularly wool or cotton) is impregnated with liquid ammonia and subsequently immersed in a dyebath without allowing complete evaporation of the ammonia [81, 82]. A residual ammonia content of about 30% is desirable. Significantly higher dye yields are claimed as the major advantage. In one example [81], cotton fabric was dyed at 60°C with a vat blue dye in the presence of caustic and hydrosulfite; the relative effect of the various pretreatments was given as follows:

Pretreatment	Relative approximate depth of shade (%)
None	100 (standard)
Liquid NH_3, evaporated in hot air	300
Liquid NH_3, squeezed to approx. 40% residual NH_3	600
Liquid NH_3, no removal of residual NH_3	600

8. LIQUID AMMONIA TREATMENT OF NONCELLULOSIC FIBERS

8.1 Wool

The cortex is the major component of wool fibers and is composed of two types of cells, ortho and para. In a typical crimped wool fiber, the cortex shows a bilateral distribution of these cells. The orthocortex lies on the convex side of the crimp wave and is more accessible to chemical reagents such as dyes, enzymes, and solvents incapable of breaking disulfide bonds than is the paracortex [83]. In liquid ammonia preferential swelling of the orthocortex can occur [84] and the resulting differential contraction of the cortical components in the fiber can lead to a bulking effect, first disclosed by Omnium de Prospective Industriel (OPI) [85]. Liquid ammonia treatment causes wool fibers to supercontract, typically 10-15%, and exhibit an increased crimp frequency. For instance, a crossbred wool yarn, treated in liquid ammonia and subsequently dried, has been claimed to exhibit bulk increases of about 35%, with optimum bulking occurring when the yarn count increases 40-60% [84]. The supercontraction, however, is reversible in water, and a net loss in fiber crimp, and thus bulk, has been seen [84]. A number of methods have been proposed to stabilize the supercontracted state.

1. Reduction of the wool fibers prior to or simultaneously with liquid ammonia treatment [86]. For instance, Hirst et al. [84] have shown that treatment of wool yarn with 1% (on weight of fiber) sodium bisulfite from aqueous solution and drying prior to treatment in liquid ammonia for 1 min resulted in a 41% increase in bulk with only an 8% decrease in yarn strength. Longer treatment times in the ammonia resulted in a progressive decrease in yarn bulk and strength, with an increase in extensibility. The same authors also claimed that stable bulk increases of up to 35% were obtained by the addition of 1-2% ammonium thioglycolate or thioglycolic acid to the ammonia bath; the reaction was, however, more difficult to control and fiber damage was severe due to excessive rupture of disulfide bonds.
2. Inclusion of cross-linking agents in the liquid ammonia bath. Hirst et al. [84] have demonstrated that a 10 min treatment of wool yarn in liquid ammonia containing 3-chloro-1,2-epoxypropane followed by curing at 110°C for 30 min yielded a yarn with 30% bulk and elasticity increases after wet relaxation. Yarn strength deteriorated only after a 30 min liquid ammonia treatment.
3. Addition of nucleophilic anions to the liquid ammonia. Vaesken et al., in a patent assigned to OPI [87], claim that although the effects of liquid ammonia treatment of wool are permanent if the immersion time is long enough (i.e., longer than 30 min), the reaction may be significantly speeded and the required time reduced to around 10 min if nucleophilic anions such as chlorides and acetates are added to the liquid ammonia, either as the salt or acid. Hirst et al. [84] have also studied the addition of ammonium chloride and acetate to the liquid ammonia, with similar results. In wool subjected to a basic medium, cystine is known to be converted to lanthionine [87, 88]; Vaesken et al. [87] claim that the effect of anions in the liquid ammonia is to catalyze the cystine-lanthionine conversion.
4. Addition of cationic agents to the liquid ammonia. Mazingue et al. [88] noted that the addition of a cationic dye (Flavine Maxilon Brillante 10GFF) as a UV tracer to the liquid ammonia solution impaired the reversibility of the supercontraction, using 15 min treatment in the liquid ammonia. It was hypothesized that the dye molecules lead to bridges in the orthocortex which stabilize the protein in its contracted form.

Pitts [89] has demonstrated that the major extent of the contraction of wool sliver in liquid ammonia occurs after only a few minutes immersion. Fine fibers were affected more than coarse fibers, with a 22 μm (64 s) sliver contracting twice as much as a 40 μm (36 s) sliver after a 15 min ammonia treatment. The fine fibers probably possess a greater proportion of orthocortex, thus their greater sensitivity. Load-extension data (e.g., stress at 15% extension, initial modulus, and breaking stress) suggested a progressive increase in the extent of hydrogen-bond rupture in the fiber with increasing immersion times in the liquid ammonia. The reductions in stress functions were greater in air than in water, which the author claimed was indicative of structural replacement of hydrogen bonds. The degree of reduction of the stress function in water, however, was great enough to indicate some irreversible structural weakening, which would be consistent with the reported lanthionine formation [88].

Wool fibers supercontracted in liquid ammonia viewed in a scanning electron microscope exhibit corrugation of the cuticle cells in the transverse direction, in agreement with a dimensionally stable cuticle forced to buckle as the cortex contracts [90]. Upon subsequent immersion in water, the corrugation disappears if the fiber has received no reducing treatment prior to immersion in liquid ammonia. The reportedly deteriorated handle of liquid ammonia treated fiber assemblies, however, does not appear to be due to the roughening of the cuticle, but is more likely due to changes in the spatial arrangement of the fibers in the yarn.

The liquid ammonia treatment of wool has been commercialized by a French company, Sitralaine S.A., specifically set up for that purpose [91]. Markets which were being considered for the wool were those where high bulk and light weight are of the most appeal, e.g., carpets, blankets, hand-knitting yarns. The three main characteristics claimed for the treated yarn are greater covering power, improved yarn elasticity (and thus improved manufacturing efficiency), and superior dye affinity. The rate of dyeing is claimed to be much more rapid than on conventional wool. The fiber, however, is also much duller. The major advantage for the process is that it would enable processing of the less desirable low-crimp wool fibers in these market areas.

By prolonged liquid ammonia treatment, Hunter et al. [92] were able to introduce some crimp into fine mohair fibers, which generally possess an axial rather than bilateral distribution of ortho and para cells; luster of the fibers was reduced and some yellowing also resulted. Supercontraction, extension at break, fiber linear density, and with-scale friction all increased with increasing treatment time, while fiber tenacity and initial modulus decreased.

8.2 Nylons

Although there appears to be no published work on the liquid ammonia treatment of nylon fibers or fabrics, the work by Zachariades and Porter [93] on plasticization of nylon 6 and 11 films should be mentioned here because of its obvious implications for textiles. Many plasticizers are available which can aid in the deformation of nylon; a disadvantage in their use, however, is that they remain in the polymer after processing and affect its properties, e.g., modulus, melting point. Ammonia was chosen for study since, due to its volatility, it had the potential for being a reversible plasticizer.

Nylon 6 and 11 ribbons of 1.5 mm thickness were treated in liquid ammonia for 5-10 hr. After removal from the treating solution they were deformed rapidly by bending, twisting, or rolling. The samples maintained their deformed shape when released from their deforming stresses after ammonia volatilization. The ammonia was shown by infrared and thermogravimetric analysis to have been absorbed by the polymer. The extent of absorption was governed by the total number of hydrogen bonds in the polymer; thus, nylon 6 absorbed 18% and nylon 11 absorbed 10% ammonia. These values were confirmed on ribbons of varying thickness (0.18-1.5 mm).

Liquid ammonia treatment also made possible the solid-state extrusion of both nylons to an extrusion draw ratio of 12 at rates and conditions not attainable with the untreated analogs. Significant increases in crystallinity and tensile modulus also resulted from liquid ammonia treatment.

9. EQUIPMENT

A detailed discussion of the equipment utilized for the liquid ammonia processing of textiles is beyond the scope of this chapter. Numerous descriptions of suggested equipment can be found, however, especially in the patent literature [46, 50, 94-101].

10. FUTURE OF LIQUID AMMONIA PROCESSING

One area which was hardly mentioned in this chapter is the processing of cotton fiber in liquid ammonia. Since it is known that liquid ammonia treatment can increase fiber cross section and elongation [13], one group of workers did evaluate its effect on the processing of low micronaire or immature fine fiber [102]. The results, however, were not promising; although cross section, fiber resiliency, fiber elongation, and drafting tenacity were increased by such treatment, increased fiber entanglement, removal and redeposition of natural waxes, and fiber rupture due to rapid expansion of liquid ammonia within the fibers during evaporation generally overcame any advantages of the treatment with regard to fiber processability.

Heap [52] has pointed out that for ammonia treatment to replace caustic treatment for mercerization, development of a water-based system for fabric treatment would be necessary. Even if this were to be done, liquid ammonia treatment probably would offer no real advantage over caustic with respect to fabric properties. Any decision to replace mercerizing equipment with ammonia equipment would probably be made on the basis of potential for automation or shorter preparation ranges, reduced effluent water-treatment requirements, and greater uniformity of treatment, all combined with a calculation of the difference in overall energy usage requirements between the two procedures.

The growth in use of liquid ammonia as a pretreatment for easy care finishing of cotton fabrics depends in large part on a continued and growing demand for high- or all-cotton content fabrics. This demand, however, might be fueled by continued optimization of finishing techniques to give products which maintain the traditional cotton absorbancy (and implied comfort characteristics) while imparting true durable press performance with minimal decrease in wear life. Several researchers believe that liquid ammonia pretreatment is only a part of the answer to achieve this goal [52, 61]. It should be combined with the latest low add-on finishing techniques, the latest chemical and mechanical finishing technology, and optimized fabric construction.

In light of the increasing number of questions being raised in various circles with regard to the safety of various textile-finishing chemicals (e.g., those containing formaldehyde), it would be a tremendous advantage if liquid ammonia technology could be developed to the extent of being able to achieve true durable press performance on cotton-containing fabrics without the necessity of applying a chemical cross-linker to the fabric.

Hudson and Cuculo [103] have reported that solutions of ammonium thiocyanate in liquid ammonia act as true solvents for cellulose. The implications of this in textile processing or textile waste reuse might be an interesting subject of study. The reversible plasticization of nylon discussed earlier [93] offers interesting prospects for fiber and textile "setting," e.g., new garment molding techniques.

It becomes clear that the future of liquid ammonia technology hinges not only on novel effects which may be accomplished through its use, but also on a complex set of factors as varied as economics, effluent requirements, chemical safety, energy supply, and new equipment developments.

REFERENCES

1. K. Hess and C. Trogus, *Ber. 68B*, 1986 (1935).
2. C. Schuerch, U.S. Patent 3,282,313 (1966).
3. R. L. Colbran, *Textiles 1* (2), 35 (1972).

4. S. A. Heap, *Text. Inst. Ind. 16,* 387 (1978).
5. *Text. Ind., 138* (9), 77 (1974).
6. J. J. Lagowski and G. A. Moczygemba, *The Chemistry of Non-Aqueous Solvents* (J. J. Lagowski, Ed.), Vol. 2, Academic Press, New York, p. 319 (1967).
7. "Ammonia," *Matheson Gas Data Book*, Matheson Company, Inc., E. Rutherford, N.J., p. 13 (1966).
8. "Anhydrous Ammonia," Chemical Safety Data Sheet SD-8, Manufacturing Chemists Assoc., Washington, D.C. (1960).
9. J. R. LeBlanc, Jr., S. Madhaven, and R. E. Porter, *Kirk-Othmer Encyclopedia of Chemical Technology*, 3rd Ed., Vol. 2, Wiley Interscience, New York, p. 470 (1978).
10. K. H. Meyer, *Natural and Synthetic High Polymers,* Interscience, New York p. 284 (1942).
11. A. J. Barry, F. C. Peterson, and A. J. King, *J. Amer. Chem. Soc. 58,* 333 (1936).
12. K. Hess and J. Gunderman, *Ber. Dtsch. Chem. Ges. 40,* 1788 (1937).
13. J. O. Warwicker, *Cell. Chem. Tech. 6,* 85 (1972).
14. G. L. Clark and E. A. Parker, *J. Phys. Chem. 41,* 777 (1937).
15. M. Lewin and L. G. Roldan, *J. Polym. Sci. C 36,* 213 (1971).
16. S. N. Pandey and R. Nair, *Text. Res. J. 45,* 648 (1975).
17. H. Z. Jung, R. R. Benerito, R. J. Berni, and D. Mitcham, *J. Appl. Polym. Sci. 21,* 1981 (1977).
18. H. J. Wellard, *J. Polym. Sci. 13,* 471 (1954).
19. J. T. Hebert, L. L. Muller, and R. M. Babin, *Microscope 23,* 139 (1975).
20. L. C. Wadsworth, J. A. Cuculo, and S. M. Hudson, *Text. Res. J. 49,* 424 (1979).
21. K. Bredereck and R. Weckmann, *Melliand Textilber. 58,* 310 (1977).
22. A. Koura and H. Schleicher, *Faserforsch. u. Textiltech. 24,* 82 (1973).
23. H. J. Marrinan and J. Mann, *J. Polym. Sci. 21,* 301 (1956).
24. J. Mann and H. J. Marrinan, *J. Polym. Sci. 32,* 357 (1958).
25. L. G. Roldan, Symposium on Modified Cellulose, Cellulose, Paper, and Textile Division, American Chemical Society, 174th National Meeting, Chicago, Ill. (August 1977).
26. J. A. Howsmon and W. A. Sisson, *Cellulose and Cellulose Derivatives* (E. Ott, Ed.), 2nd Ed., Interscience, New York, p. 241 (1954).
27. J. Mercer, British Patent 13,296 (1850).
28. J. O. Warwicker, R. Jeffries, R. L. Colbran, and R. N. Robinson, *A Review of the Literature on the Effect of Caustic*

Soda and Other Swelling Agents on the Fine Structure of Cotton, Shirley Institute Pamphlet No. 93, Cotton, Silk and Man-Made Fibres Research Association, Manchester, England (1966).
29. D. Bechter, D. Fiebig, and S. A. Heap, *Textilveredlung 9,* 265 (1974).
30. *Textile Month* October 1969, p. 82.
31. J. & P. Coats Ltd., British Patent 1,135,417 (1968).
32. R. M. Gailey, *Textilveredlung 7,* 789 (1972).
33. *Daily News Record* July 5, 1972, p. 2.
34. J. Aitken and W. E. Graham (to J. & P. Coats Ltd.), U.S. Patent 4,106,902 (1978).
35. OPI Cryochimie, British Patent 1,477,707 (1977).
36. T. V. Ivanova, G. I. Vinogradova, B. N. Mel'nikov, N. P. Bazhanova, and E. A. Osminin, *Tekstil'naya Promyshlennost' 38,* 59 (1978).
37. M. L. Nelson, C. B. Hassenboehler, Jr., F. R. Andrews, and A. R. Markezich, *Text. Res. J. 46,* 872 (1976).
38. F. R. Andrews, C. B. Hassenboehler, Jr., M. L. Nelson, and A. R. Markezich, *Text. Res. J. 47,* 670 (1977).
39. M. A. Rousselle, M. L. Nelson, C. B. Hassenboehler, Jr., and D. C. Lengendre, *Text. Res. J. 46,* 304 (1976).
40. M. A. Rousselle and M. L. Nelson, *Text. Res. J. 46,* 648 (1976).
41. K. Schwertassek and V. Hochman, *Melliand Textilber. 55,* 544 (1974).
42. K. Bredereck, *Textilveredlung 13,* 270 (1978).
43. M. B. Roberts and L. Hunter, *SAWTRI Bull. 12* (March), 43 (1978).
44. *Textile Month* January 1971, p. 48.
45. *Melliand Textilber. 57,* 684 (1976).
46. Cluett, Peabody & Co., Inc., W. S. Troope, and J. Lawrence, British Patent 1,365,706 (1974).
47. Sentralinstitutt for Industriell Forskning, British Patent 1,084,612 (1967).
48. *Textiles 7* (2), 41 (1978).
49. *Text. Manufacturer 77* (4), 36 (1977).
50. J. B. Carpenter, Jr. (to Burlington Industries, Inc.), U.S. Patent 4,051,699 (1977).
51. W. S. Troope, *Text. Res. J. 50,* 162 (1980).
52. S. A. Heap, *Text. Inst. Ind. 16,* 387 (1978).
53. K. Bredereck, *Textilveredlung 13,* 498 (1978).
54. K. Bredereck and S. A. Heap, *Textilveredlung 9,* 251 (1974).
55. T. A. Calamari, Jr., S. P. Schreiber, A. S. Cooper, Jr., and W. A. Reeves, *Text. Chem. Colorist 3,* 234 (1971).

56. K. Bredereck and R. Weckmann, *Melliand Textilber.* *59,* 137 (1978).
57. K. Bredereck, *Textilveredlung 13,* 270 (1978).
58. K. Bredereck, *Melliand Textilber. Intl. 59,* 648 (1978).
59. K. Bredereck, *Melliand Textilber. 60,* 1027 (1979).
60. F. A. Barkhuysen, SAWTRI Tech. Report No. 289 (1976).
61. B. W. Jones, J. D. Turner, and D. O. Luparello, *Text. Res. J. 50,* 165 (1980).
62. J. D. Turner, J. N. Etters, W. L. Mauldin, A. R. Urbanik, J. H. Wood, and M. A. Zentner, *Text. Chem. Colorist 9,* 250 (1977).
63. S. L. Vail, G. B. Verburg, and T. A. Calamari, Jr., *Text. Chem. Colorist 7,* 175 (1975).
64. C. J. Gogek, W. F. Olds, E. I. Valko, and E. S. Shanley, *Text. Res. J. 39,* 543 (1969).
65. H. Z. Jung, R. J. Berni, R. R. Benerito, and M. W. Pilkington, *J. Appl. Polym. Sci. 20,* 1905 (1976).
66. M. Lewin, R. O. Rau, and S. B. Sello, *Text. Res. J. 44,* 680 (1974).
67. I. V. Bulavkina, M. N. Kirillova, and B. N. Mel'nikov, *Tekstil'naya Promyshlennost' 36* (6), 67 (1976).
68. E. Böhmer and B. Ormestad, *Proceedings of the Symposium on Man-Made Polymers in Papermaking,* Helsinki, p. 273 (1972).
69. I. Lambrinou, *Melliand Textilber.* 56, 277 (1975).
70. K. Bredereck and R. Pfundtner, *Textilveredlung 10,* 92 (1975).
71. T. A. Calamari, Jr., S. P. Schreiber, and W. A. Reeves, *Text. Chem. Colorist 7,* 65 (1975).
72. T. A. Calamari, Jr., S. P. Schreiber, L. W. Mazzeno, Jr., and W. A. Reeves, Paper presented at the 12th Cotton Utilization Conference, New Orleans, La. (May 1972).
73. F. A. Barkhuysen, SAWTRI Tech. Report No. 406 (1978).
74. F. A. Barkhuysen, SAWTRI Tech. Report No. 427 (1978).
75. F. A. Barkhuysen, *SAWTRI Bull. 12* (3), 16 (1978).
76. F. A. Barkhuysen, SAWTRI Tech. Report No. 447 (1979).
77. R. H. Wade and J. J. Creely, Paper presented at the 12th Cotton Utilization Research Conference, New Orleans, La. (May 1972).
78. R. H. Wade, *Sources and Resources 7* (5), 21 (1974).
79. J. P. Tratnyek, U.S. Patent 3,666,398 (1972).
80. S. M. Kane, *Amer. Dyestuff Reptr.* May 1973, p. 27.
81. J. P. Dalle (to Omnium de Prospective Industrielle, S.A.) U.S. Patent 3,892,521 (1975).
82. OPI Cryochimie, A. DeClercq, and P. DeClercq, British Patent 1,454,890 (1976).
83. R. H. Peters, *Textile Chemistry,* Vol. I. Elsevier, Amsterdam (1963), pp. 269-270.

84. L. Hirst, J. M. D. Pitts, and R. Umehara, *Proceedings of the Fifth Intl. Wool Text. Res. Conference* (K. Ziegler, Ed.), Vol. 4, German Wool Res. Institute, Aachen, Germany (1976).
85. OPI Cryochimie, French Patent 72,08875 (1972).
86. IWS Nominee Co., Ltd., British Patent Appl. 19691/75 (1975).
87. M. J. Vaesken, A. E. Bultez, and J. P. Dalle (to OPI Cryochimie, S. A.), U.S. Patent 4,030,883 (1977).
88. G. Mazingue, M. Bauters, J. Jacquemart, and R. Hagege, *Bull. Scient. ITF 4* (13), 77 (1975).
89. J. M. D. Pitts, *J. Text Inst. 67,* 12 (1976).
90. B. N. Hermiston and J. M. D. Pitts, *J. Text. Inst. 67,* 72 (1976).
91. P. Lennox-Kerr, *Text. Ind. 140* (5), 65 (1976).
92. L. Hunter, S. Smuts, and F. A. Barkhuysen, SAWTRI Tech. Report No. 372 (1977).
93. A. E. Zachariades and R. S. Porter, *J. Appl. Polym. Sci. 24,* 1371 (1979).
94. Cluett, Peabody & Co., Inc., J. Lawrence, and W. S. Troope, British Patent 1,419,921 (1975).
95. Cluett, Peabody & Co., Inc., and J. Lawrence, U.S. Patent 3.939,576 (1976).
96. Cluett, Peabody & Co., Inc., and J. Lawrence, U.S. Patent 4,074,969 (1978).
97. OPI Cryochimie, S. A., British Patent 1,486,160 (1977).
98. Tedeco Textile Development Co., Netherlands Patent 7006607 (1970).
99. E. C. Hanekom and F. A. Barkhuysen, SAWTRI Tech. Report No. 277 (1975).
100. Cluett, Peabody & Co., Inc., British Patent 1,528,409 (1978).
101. J. & P. Coats Ltd. and R. M. Gailey, British Patent 1,519,035 (1978).
102. C. O. Graham, P. L. Rhodes, T. A. Calamari, and R. J. Harper, *Text. Res. J. 46,* 158 (1976).
103. S. M. Hudson and J. A. Cuculo, Paper presented at the 179th National Meeting, American Chemical Society, Houston, Tex. (March 1980).

5

RAW-WOOL SCOURING, WOOL GREASE RECOVERY, AND SCOURING WASTEWATER DISPOSAL

GEORGE F. WOOD* Commonwealth Scientific and Industrial Research Organization, Division of Textile Industry, Belmont, Victoria, Australia

1. Introduction 206

1.1 Properties of raw wool 206
1.2 Prescouring operations 209
1.3 Felting in scouring 210
1.4 Scouring as a multistage countercurrent extraction process 211

2. Detergent Scouring 213

2.1 Water supply 213
2.2 Mechanical systems used in scouring 213
2.3 Drying 219
2.4 Types of detergent systems 219
2.5 Suint scouring 223
2.6 Transference of contaminants from wool to liquor 223
2.7 Liquor systems 225
2.8 Quality control and product testing 231

3. Wool Grease Recovery 233

3.1 The stability of wool grease emulsions 233
3.2 Recovery systems 235
3.3 Centrifuging 236
3.4 Wool grease refining 239

4. Wastewater Disposal 241

4.1 General considerations 241
4.2 Types of processes 244
4.3 Processes used in industry 246

*Now retired.

5. Energy Conservation 249

6. Solvent Scouring 250

6.1 General considerations 250
6.2 Discontinued solvent scouring processes 251
6.3 The SOVER process 252

7. Other Methods of Cleaning Raw Wool 253

7.1 Chilling and dusting 253
7.2 Treatment with powders 253
7.3 Processing wool without scouring 254

References 254

1. INTRODUCTION

1.1 Properties of Raw Wool

Raw wool is the natural coat of the sheep, removed by shearing. In addition to the wool fiber it usually contains wool wax (the excretion of the sebaceous gland), suint (the excretion of the sudoriferous gland), inorganic dirt (clay, sand), organic dirt (urine, feces, organic soil components), vegetable matter, and water.

Fleece Structure

Wool consists of two types of fibers, primary and secondary. On the skin of the sheep, they occur in groups of three primary fibers associated with several secondary fibers [1]. The ratio of the number of primary fibers to the number of secondary fibers averages from about 25 for merino sheep to about 5 for coarse-wooled breeds. Every fiber has an associated sebaceous gland, but only primary fibers have a sudoriferous gland. Therefore, the ratio of wool wax to suint is high for fine-wooled sheep and low for coarse-wooled sheep, a fact with important implications in scouring.

As the fleece grows, the fibers form into bundles called staples, which retain their identity apparently through the bonding effect of a "cement" formed at the tip by a mixture of wool wax, suint, soil, and water. In some cases the tip can be extremely hard and cohesive, and is removable in scouring only with great difficulty. The number of fibers in a staple may vary from about 5000 to perhaps 50,000.

Staple formation is important in scouring because the parallel fibers in the staples support each other and reduce relative movement that could lead to felting.

Distribution of Contaminants

Little is known about the fiber-by-fiber distribution of contaminants in the fleece, but gross distribution follows a consistent pattern. Adventitious contaminants such as soil and vegetable matter are naturally concentrated on the outside of the fleece, i.e., at the tips of the staples. Vegetable contamination is often particularly heavy on the belly and legs. Feces and urine contaminate principally the breech, the belly, and the legs. Wax and suint are distributed throughout the fleece, but wax concentrations are usually higher at the root end of a staple than at the tip, and suint, being water soluble, tends to wash off the surface and migrate to the deeper layers of the fleece and to the belly.

Despite the fact that in fine wool only 1 fiber in 25 may have an associated suint gland, available evidence suggests that the suint migrates extensively and is probably present on most fibers. When a single fiber of raw wool is viewed under the microscope, it is very difficult to distinguish one contaminant from another. However, it seems that the suint and grease are distributed along the fiber fairly evenly, with an occasional lump. It is highly probable that the contaminants are partly mixed or emulsified and that these mixtures later contribute to the complex structure of the scouring liquor.

Wool Wax

Whether it exists as pure wax or as a mixture with other contaminants, the wool wax is partially degraded during its exposure on the sheep, and for the purposes of scouring technology is usually regarded as comprising *oxidized wax* and *unoxidized wax*. The oxidized wax is located on the tip half of the growing fiber and has had a long exposure to air, sunlight, and water. The unoxidized wax is located on the root half of the fiber and has had only a relatively short exposure after excretion from the gland. The two fractions are present in similar proportions [2].

There is no way of deciding unequivocally whether a given wax sample is oxidized or unoxidized. In practice the distinction is made according to the behavior of the wax in the scouring situation, and is really a function of its surface-chemical properties [3]. It is quite likely that a sample of wax regarded as oxidized wax may be a mixture of a small proportion of highly surface-active material with a large proportion of inert material that on its own would be regarded as unoxidized.

Both oxidized and unoxidized wool waxes have melting points around 40°C. Since removal of *solid* wax from wool by detergent solutions is slow and difficult, 40°C is the lowest temperature at which detergent scouring liquors are effective for removing wax.

The chemical composition of wool wax is not important in the scouring context and will not be dealt with here. The interested reader is referred to the standard work on the subject [4] and four recent reviews [5-8].

The wool wax recovered during the scouring process is usually called wool grease. Refined wool grease is called lanolin. However, there are no universally accepted definitions of these terms.

Suint

Suint is water soluble; its removal from raw wool in scouring is therefore easy. Its solutions are alkaline because it contains a high percentage of dissociable potassium salts of organic acids. This alkalinity and the presence of potassium ions have important effects on the overall behavior of scouring liquors [9].

Inorganic Dirt

Sand and clay particles in raw wool vary from large grains to colloid-size particles. In the scouring process, sand grains have to be dislodged by mechanical action. Clay particles, because of their well-known surface activity, probably would not exist alone, but would always be associated with soil organic matter, wool wax, and suint. They are removed by detergency mechanisms and contribute to the distinctive properties of scouring liquors.

Organic Dirt

Serious investigation of the dirt component of scouring liquors (i.e., the insoluble, nonlipid component) started only recently [10]. The organic part of this dirt has become of vital importance since the introduction of scouring techniques using very small volumes of water. In traditional scouring small quantities of residual organic dirt are removed by extensive rinsing; when rinsing is drastically reduced the scoured wool often has an unacceptable gray appearance.

Organic dirt originates in fecal matter, in dead cells (*epithelial debris*) [11], and possibly in organic soil components. The reasons for the persistence of these particles through the ordinary detergent process and for the difficulty in removing them from the wool fibers are not yet understood.

Vegetable Matter

Vegetable matter is not removed to a significant extent in the scouring process, but it has an indirect effect in that heavy vegetable contamination can bind the fibers together, thus hindering the penetration of the wool by detergent solutions and hence the removal of contaminants.

Raw Wool Composition

Water is the least variable component of raw wool, and is usually within the range of 9-12%. The other components may vary quite widely, but the relationships between fiber diameter, grease content, and suint content noted under Fleece Structure above apply in a broad fashion [12, 13].

The compositions given in Table 5.1 illustrate typical wools of different fiber diameters. Some idea of the composition of other wools with the same fiber diameter but lower yield may be obtained by supposing that the lower yield results simply from the addition of extra dirt or vegetable matter.

Skin Wools

Skin wools, removed from sheepskins by chemical or biological depilatory processes, lose a fair proportion of their contaminants during these processes, so have a higher yield than the corresponding raw wools. Scouring of chemically depilated skin wools requires special conditions because of quantities of calcium salts remaining in the wool.

1.2 Prescouring Operations

Prescouring operations with a potential effect on the scouring process are sheep branding, baling, sorting, opening, and feeding.

Sheep Branding

The widespread use of scourable brands [14] has largely eliminated the brand problem, but it should be noted that heavy brands are not completely removed in neutral scours or in scours with very short residence times.

Table 5.1 Illustrative Compositions of Different Wools

Fiber diameter (μm)	Yield (%)	Clean dry fiber (%)	Water (%)	Grease (%)	Suint (%)	Dirt (%)	V.M. (%)
23	63	54	10	15	5	15	1
23	54	45	10	8	4	30	3
28	72	63	11	9	12	4	1
35	75	66	12	5	10	7	-
45	87	76	12	5	7	-	-

Baling

Wool from a modern dense bale must be allowed to expand before opening. Warming the bales in a hot room or by microwave heating are satisfactory procedures.

Sorting

Sorting is labor intensive and has become rare today. Where practiced, it serves as a very gentle opening operation and may obviate the need for more vigorous, potentially damaging opening.

Opening

Opening serves two purposes: it removes a proportion of solid contaminants, mainly dirt, and it leaves the wool in a more open condition by breaking down clumps and staples bound together by vegetable matter, mud, or hard tip, or felted in cots [15]. Such clumps cannot be penetrated by scouring liquor, and they trap contaminants within themselves.

Since removal of dirt in an opener is much cheaper than removing it in a scour, and since better opened wool gives a cleaner product, scourers have a natural desire to open the wool to the maximum possible extent. However, opening also leads to more felting in the scour (see Sec. 1.3), so a balance must be achieved. In general, carding wools are given severe opening treatments and combing wools are treated more gently.

Feeding

The standard feeding hopper, comprising a spiked lattice, a comb to remove excess wool from the spikes, and a beater to transfer the remaining wool to the scouring machine, also serves as a mild opener. Even today, a significant proportion of combing wools receives no opening other than that provided by the feeding hopper.

The most important function of the feeding hopper is to deliver the wool to the scouring machine in a uniform layer. Studies have shown that standard hoppers are incapable of achieving the uniformity required by modern scouring techniques, and the industry is looking toward weigh belt feeders and other improved feeding mechanisms [16]. Uniform feeding allows the full production potential of the scouring machine to be utilized, and of course helps eliminate nonuniform scouring.

1.3 Felting in Scouring

Although it has been shown that felted wool gives many disadvantages (e.g., lower yarn strength) in the woolen system, no regard is usually paid to felting in the scouring of carding wools. However, topmakers have always been aware of the penalties attached to felting, and conversely, the benefits to be gained by reducing felting in combing wools. If top is worth $1 per kg more than noil, a reduction of 10% in the quantity of noil produced in combing is worth $0.007 per kg, or $35,000 per year to a mill combing 5 million kg/year.

In the two decades after 1950 considerable experimentation took place to reduce felting in scouring. As a result of this work, several new scouring machines were developed to the industrial stage, including the SOVER solvent process [17], the UNISEARCH [18] and the CSIRO [19, 20] jet machines, the WIRA improved scouring machine [21], and the Fleissner machine [22]. Of these, only the Fleissner machine achieved wide industrial acceptance.

In practice it was found that with all these processes a reduction in felting was achieved at the expense of cleanliness. All attempts to break this nexus failed, and it is now accepted that if a scouring machine is well operated, no worthwhile reduction in felting can be achieved without some reduction in cleanliness.

There was a time when a slight reduction in the cleanliness of scoured wool had only a marginal effect on the economics of topmaking, and the cost balance between gains due to reduced felting and losses due to reduced cleanliness was in favor of reduced felting. Owing to changes in industrial practice, the balance has now swung the other way, and cleanliness has become of overriding importance. The conventional scouring machine in one of its modern forms is consequently the best type of machine to use for combing wools.

Wool opening may be an important contributor to felting in the scouring process. If staples in the opened wool remain more or less intact, little effect on felting will be seen, but if the staples are intensely subdivided, serious felting may result. For all combing wools, opening should be the minimum consistent with the degree of cleanliness that must be achieved.

Felting is also greatly influenced by mechanical conditions in the scouring machine. Vigorous agitation in the scouring bowl caused by high rake speeds or turbulence in the liquor can cause serious felting. Similarly, undue movement of wet wool during transport between bowls or from the last bowl to the dryer is undesirable.

Felting in conventional scouring bowls is minimized by using high wool feed rates. In the thick layer of wool in the bowls, adjacent fibers and staples provide mutual support and inhibit felting movements.

1.4 Scouring as a Multistage Countercurrent Extraction Process

Scouring may be regarded as an extraction process in which the contaminants are extracted from wool by a detergent solution. The most efficient way of carrying out such a process is as a continuous countercurrent extraction (see Sec. 2.7).

The evolution of scouring methods over the last decade has been toward the ideal of a steady-state, continuous process. However, in order to achieve a steady state, certain conditions must be met, one of which is that the liquor in each stage must be homogeneous.

Failure to achieve homogeneity is the main reason that industrial scouring machines have not been operable as steady-state processes. Dirt is the component that causes heterogeneity; it settles from the liquor in the scouring bowls, eventually building up to such an extent that the bowls have to be washed out.

Many devices have been used in attempts to remove dirt from the bowls and so extend the period between successive washouts. The so-called self-cleaning bowl has a hopper bottom from which sludge is removed continuously or intermittently by a screw conveyor. Liquors have been circulated continuously through sludge-discharge centrifuges or through settling tanks.

These methods fail because settlement of dirt takes place not only into the bottom of the bowl (from which it can be removed) but also onto the perforated false bottom and the sloping sides of the hopper bottom. The maintenance at all points in the bowl of liquor velocities high enough to prevent settling would possibly be in conflict with the essential requirement of gentle motion to prevent felting.

Nevertheless, substantial progress has been made toward steady-state operation. Several industrial plants regularly run for 4 weeks between washouts, and intervals of 1 week are becoming common. Very dirty wools, however, are still run with intervals as low as 8 hr between washouts.

One of the factors encouraging the trend toward steady-state operation is the rapidly increasing cost of effluent disposal. The closer an operation approaches steady-state conditions, the lower the water consumption, and hence the lower the quantity of effluent to be treated. Also, it is a characteristic of steady-state systems that the volume of extractant required to attain a given extraction efficiency can be reduced by increasing the number of extraction stages. In scouring terms, this means that water consumption can be reduced by increasing the number of scouring bowls. Recently, attempts have been made to move in this direction by introducing scouring systems in which one or more bowls in a conventional scouring machine are replaced by a greater number of small bowls [23].

Whatever the type of scouring machine employed, the scouring manager should be aware that the closer the operation moves toward steady-state operation, the more efficient it will become, and that influences prejudicial to this type of operation (e.g., unregulated discharges from various parts of the machine, wool feed variations, flow-rate fluctuations) inevitably reduce efficiency.

2. DETERGENT SCOURING

2.1 Water Supply

When soap was universally used for scouring, a good soft water supply was essential. Although synthetic detergents are as effective in hard water as in soft, little if any relaxation in water quality standards can be tolerated when detergents are used for wool scouring.

Of particular importance are traces of iron, manganese, and other heavy metals. Even levels lower than 1 mg/liter can be sorbed into wool fibers in sufficient quantity to cause discoloration or to adversely affect the dyeing of the wool. Calcium has also been shown to be sorbed under appropriate conditions [24], giving undesirable effects. Traditional water quality standards should accordingly be retained, irrespective of the detergent system used.

2.2 Mechanical Systems Used in Scouring

The machine used in scouring must have a means for bringing the wool into contact with the scouring liquor, and must also have conveying elements to move the wool through the various washing stages. The contact of the wool with the liquor may be gentle or vigorous. A vigorous action gives better scouring but tends to felt the wool. The conveying elements may also produce felting.

Traditional Scouring Machines

In the traditional machine wool is propelled through long bowls of liquor by means of long-fingered rakes. This method was originally developed because of its capability for cleaning wool well without excessive felting. The elements of a traditional bowl are shown in Fig. 5.1. At the wool entry point, a *dunker* is usually included to submerge the wool (otherwise it has a tendency to float). The wool is then raked through the bowl and conveyed by means of a transfer mechanism to squeezing rollers, where as much liquor as possible is removed before the wool enters the next bowl. The bowl has a false

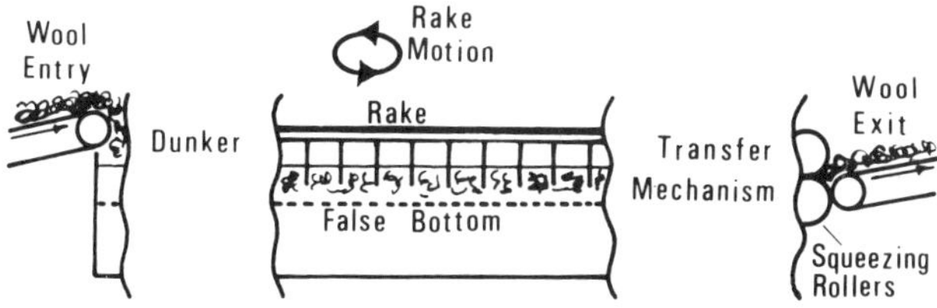

Figure 5.1 The elements of a conventional scouring bowl.

bottom consisting of a perforated plate which allows dirt removed from the wool to fall through the perforations while preventing fibers from passing. The plate also prevents wool from entering circulating streams of liquor that are drawn from the bowl. When the wool leaves the last scouring bowl it is passed through a dryer. All the mechanical elements in a scouring train—dunker, rakes, transfer mechanism, and squeezing rollers—may have effects on scouring efficiency and felting, and are discussed in this section. Liquor circulation, which also affects scouring and felting, is discussed later.

Dunkers. Usually dunkers are perforated, inverted baskets attached to the frame of the rakes. The vigor of their action is dependent on their shape and degree of perforation as well as on rake speed. Sometimes separate dunkers (for example, ribbed rotating drums) may be used.

Rakes. A *harrow motion* is preferred for combing wools, where the gentlest action is required. The harrow consists of a frame bearing several transverse rows of straight prongs, each row covering the full width of the bowl. The motion is ideally a square one, in which the prongs are pushed vertically downward into the liquor, moved forward in a horizontal line, lifted vertically out of the liquor, and returned horizontally backward to the starting position. Rake speeds are low, on the order of 10 revolutions per min.

Rake motions are used for carding wools. The prongs of the rakes are often curved instead of straight, and the motion is circular rather than square. The rakes are in three longitudinal rows moving 120° out of phase with each other, and this permits the rakes to move much faster than a harrow does. (A fast harrow motion would produce unmanageable waves in the bowl.) The greater vigor of a rake motion is due entirely to the speed. Slow rake motions can be used quite satisfactorily for combing wools without felting them.

Transfer mechanisms. The transfer of wool carrying several times its own weight of water into the squeezing rollers can under some circumstances be extremely difficult. The problem arises particularly with short wools and greasy or foamy liquors; the slippery liquor prevents the rollers from gripping the wool and pulling it into the nip. This difficulty is more prevalent in modern scours with low water usage, since the liquors are more foamy and slippery than when larger quantities of water are used.

The washplate is by far the most common transfer mechanism, and is shown diagrammatically in Fig. 5.2. Its main advantages are that it tends to even out irregularities in the wool flow arising from the raking motion, and it is less susceptible to mechanical failure since it has no moving parts.

Conveyors are used successfully with short wools. Longer wools tend to jam between the lower squeezing roller and the conveyor.

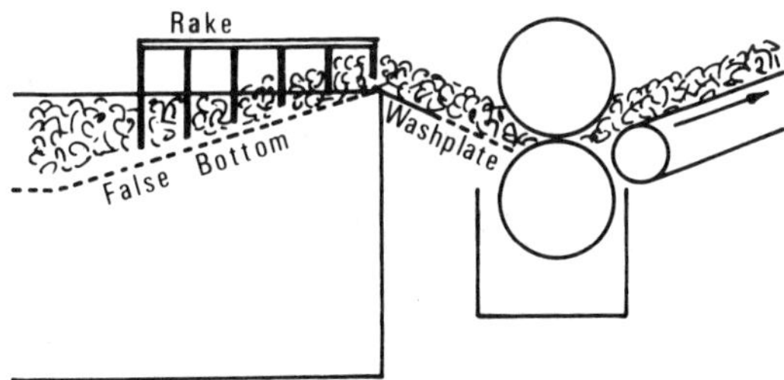

Figure 5.2 The washplate. Scouring liquor overflows from the bowl and washes the wool down the plate into the squeezing rollers. The plate is perforated at its lower end to assist drainage of liquor from the wool.

The Belgian lift is still frequently seen in European mills, but may be regarded as obsolete.

Wool transfer seldom fails totally. Rather, wool is held up temporarily, across only a part of the width. Such holdups result in the formation of lumps, which tend to grow larger as the wool progresses through the machine. The lumps are subjected to much agitation as they hesitate at the squeezing roller entrance, and severe felting may occur. The felted lumps cannot be scoured or dried properly, and product quality is unsatisfactory. Productivity also falls, since production rates must be reduced in an effort to cope with the lumps. Such effects of wool transfer are sometimes overlooked, and this aspect of scouring is often not given the attention it merits.

Squeezing. Squeezing rollers are universally used for dewatering wool. Other mechanisms have occasionally been suggested or tried [25], but none has achieved industrial acceptance. The best possible squeeze reduces the water content to about 55% calculated on the dry fiber weight, i.e., about 25% free water and 30% internal water bound into the fiber structure. Mill squeezing rollers usually give 60-80%.

Squeezing efficiency is determined mainly by the pressure applied to the rollers, but it also depends on the material of construction. The bottom roller is always metal—formerly cast iron, currently stainless steel—and is usually knurled or otherwise roughened to assist in gripping the wool. The top roller may be covered with a resilient material such as rubber or polyurethane, wrapped (lapped) with several layers of wool or other fibrous material, or constructed in some other way such as from compressed fiber disks or resin-bonded fiber.

The most efficient squeeze is given by top roller coverings that are both resilient and absorptive, namely fibrous lapping. However, these are also the most expensive to install and maintain.

The function of the final squeeze is to remove as much water as possible before drying; mechanical removal is much cheaper than thermal removal. Hence a lapped top roller is usually used in this position. Since squeezing efficiency is greater with warm water than with cold water, there is a case for having a warm rinse rather than a cold one as the final bowl [26].

The function of the intermediate squeezing rollers is to minimize the carry-over of liquor from one bowl to the next. Where water usage is relatively high (say 15-20 liters per kg of greasy wool), the amount of carry-over has little effect on the overall scouring efficiency, and traditionally the intermediate squeezing rollers may have lower efficiency than the final ones. However, in modern plants with water usage less than 5 liters per kg, the carry-over has an important effect on overall scouring effectiveness, and all squeezing rollers are required to be highly efficient.

Besides their primary use in removing water from wool, squeezing rollers also serve the important function of breaking down hard tip, mud, dag, and other lumpy impurities, thus facilitating their removal in the scouring bowl. The action of squeezing itself has no appreciable felting action on the wool.

Unconventional Machines

As previously stated, little attention is paid to felting in the scouring of carding wools, which are invariably scoured with a vigorous action to produce the cleanest possible product. Much the same applies to coarse wools such as carpet wools. With the possible exception of the Fleissner machine, the nontraditional mechanisms described in this section are therefore applicable only to combing wools, since they are all designed to reduce felting. An interesting review of unconventional machines has been given by Poncelet [27].

The Fleissner machine. In the Fleissner scouring bowl (Fig. 5.3) wool is conveyed through the bowl by a series of suction drums. Liquor is circulated by pumps through the perforated surface of the drum from

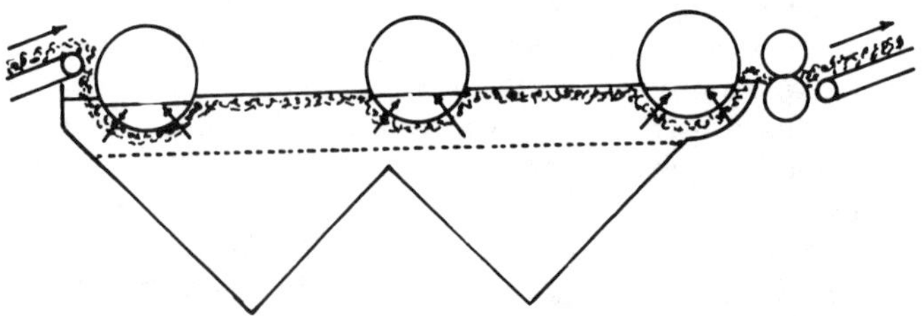

Figure 5.3 Fleissner scouring bowl. The flow of liquor through the wool and into the three suction drums is indicated by arrows.

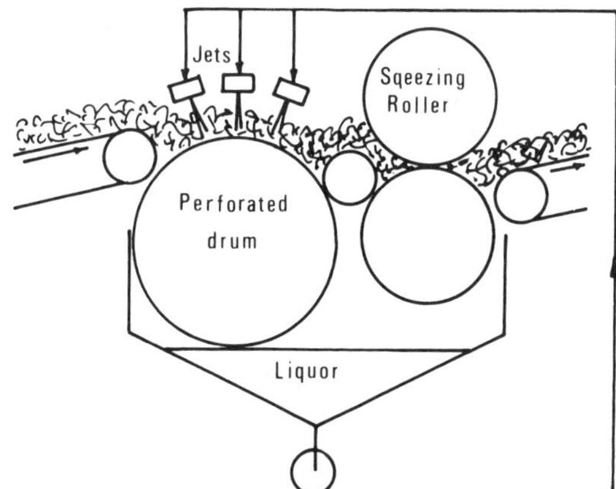

Figure 5.4 Jet scouring bowl.

outside to inside, thus holding the wool layer against the surface while the liquor is drawn through it. The wool detaches itself from the drum near the liquor surface and floats through the liquor to the next drum. At the end of the bowl the wool is conveyed to a squeezing set.

The scouring action of a Fleissner machine is very gentle, and felting in the scoured wool is significantly less than in a conventional machine. As always, the reduced felting is accompanied by lower wool cleanliness; to minimize this effect Fleissner scours are usually preceded by a vigorous opening section. Even so, Fleissner scoured wools tend to be slightly dirtier than conventionally scoured wools, a limitation that could in the long run determine the usage of the machines. They are significantly more expensive than conventional machines.

Jet scouring machines. A number of variations of jet scouring have operated in mills for various lengths of time during the last 20 years. In the most common arrangement (Fig. 5.4) the wool travels on a conveyor belt or a perforated drum below low-pressure jets or sprays of scouring liquor [19, 20]. Because the jet velocity is low, little movement of the wool occurs and less felting takes place than in a conventional scouring bowl. (However, severe felting is possible if high jet velocities are used, so jet scouring is not automatically a low-felting procedure.)

In other arrangements, the wool may be constrained between two conveyor belts or between a conveyor belt and a perforated drum, and the jetting may take place while the wool is submerged [18]. Several mills have used arrangements where one or more conventional bowls are followed by jetting bowls.

In the Lo-flo process for pollution control [23], to be further discussed later, washplate bowls are used in which the wool is washed down a perforated plate into the squeezing rollers by jets or sprays of scouring liquor (Fig. 5.5). These bowls are designed for liquor handling rather than scouring, and their scouring efficiency is low.

Because of the connection between felting and wool cleanliness noted previously, it unlikely that jet machines will be much used in the future, though for special purposes jet bowls may be incorporated into otherwise conventional machines.

Although the CSIRO jet scouring machines of the 1960s did not achieve wide industrial acceptance, they were pioneering machines in their use of many principles that are considered important in modern scouring systems. They used small bowl sizes, full countercurrent flow controlled by a single discharge valve, enclosed bowls to reduce heat losses, indirect heaters rather than live-steam injection, liquor storage tanks separate from the bowls to facilitate solids removal, bowls designed to prevent sludge accumulation, and other features that are given prominence in some of the systems being promoted today.

The Hydronamic scour. This system [28] involved gilling the greasy wool to deliver it to the scouring machine as a very uniform web with a high degree of fiber parallelization, and then scouring the web while it was constrained between two moving belts. It was not developed beyond the pilot-plant stage.

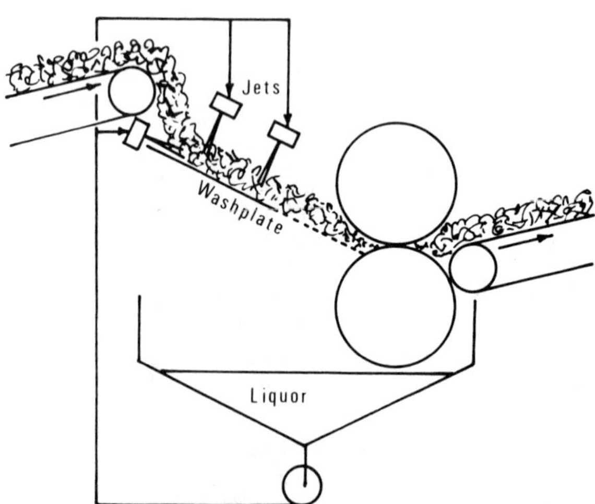

Figure 5.5 Washplate bowl as used in the Lo-flo process. This is the shortest possible bowl, but has poor scouring efficiency.

Pipe scouring. In this system [29] the scouring bowl is replaced by a pipe through which wool and liquor flow concurrently, the flow being induced by injected air. It was not developed beyond the pilot-plant stage.

2.3 Drying

In loose wool drying, the most important single factor is the uniformity of the wool layer. A uniform layer gives more even drying, more efficient heat utilization, and greater productive capacity than a nonuniform layer. For this reason the provision of a feed hopper between the final squeezing rollers and the dryer is becoming increasingly common. Although there is evidence that the feed hopper may cause some felting, this additional felting does not appear to be significant in normal mill processing.

Another important factor in drying is the velocity of air through the wool mass. Much higher velocities are achieved in drum dryers than in conveyor dryers, and this is the main reason for the great increase in the popularity of drum dryers over recent years. Dryers in which settings for operating temperature, residence time of the wool, and exhaust rate of air can be readily adjusted are easy to run optimally.

A dryer based on a new design principle was being tested as a commercial prototype at the time of writing [30].

2.4 Types of Detergent Systems

Detergents (using the word in its broadest sense) used in wool scouring are soap, synthetic nonionic detergents, and synthetic anionic detergents. Soap must always be accompanied by an alkali, soda ash being the almost universal choice. Synthetic detergents may be used alone or with alkaline or neutral builders.

If used correctly, all the commonly used materials give excellent results, and the choice between them depends on economic factors. These include not only the cost of the detergent itself, but also the cost of dissolving it (where this is necessary) and dispensing it to the scouring machine. Costs may also be influenced by effluent disposal conditions—the detergent may have to be soft (biodegradable) or have other specified properties.

The remainder of this section is taken up by a discussion of detergent properties in relation to the scouring process and the properties of the wool product. The actual method of using detergents is covered in Sec. 2.7 on liquor systems.

Soap

Soap is the most difficult detergent to use. Dissolving it in hot water is difficult, and the solutions gel if allowed to cool in storage tanks or pipes. (Some plants today buy their soap as hot solutions, delivered by tank truck.) Soap must always be accompanied by alkali, usually soda ash, which must also be dissolved before use.

If the pH of a soap scouring liquor is allowed to fall below about 8.0, hydrolysis occurs and the free fatty acid produced will exhaust onto the wool, giving it an unacceptably high solvent extract and usually a poor color. On the other hand, if the pH of the later bowls becomes too high, the wool will be damaged and yellowed in the dryer. Control of the process is therefore quite critical, and it is advisable to have at least two rinse bowls following the soap-containing bowls.

Soaps used for scouring are usually sodium (sometimes potassium) soaps of saturated and unsaturated fatty acids. A typical product has a total fatty matter content of 80%, the titer of the fatty acids being 35°C.

The use of soap for wool scouring diminished greatly between 1950 and 1970, its place being taken by nonionics which were cheaper and easier to handle. The rapid increase in detergent prices in the later part of the 1970s, together with the fact that soap is biodegradable whereas the most widely used nonionics were not, prompted many scourers to look at soap again. It is possible that soap will again be widely used in the future.

Nonionic Detergents

When nonionic detergents were first being tried by the industry, many claims were made for them and accusations leveled against them [31-34]. Some advantages claimed were that the scoured wool was whiter, loftier, and stronger, and that it processed better. A disadvantage was supposed to be that the detergent overscoured the wool, leaving it brittle and so causing fiber breakage in processing. Wool grease recovered from detergent scouring was claimed by different workers to be higher or lower in yield and better or worse in quality.

Most of these claims arose because comparisons were being made with poorly controlled soap scouring, or because optimum conditions for nonionic scouring had not been worked out, or because the wrong type of nonionic detergent was being used—different types differ very widely in effectiveness.

As time went by, a single type of nonionic detergent came to dominate the scouring industry. This was an octyl- or nonylphenol ethoxylate with 8-10 ethylene oxide units, which reigned supreme until biodegradability requirements severely curtailed its use. None of the products that succeeded it have attained quite the same scouring efficiency (in terms of weight of detergent used per unit weight of wool scoured), the closest being some secondary alcohol ethoxylates in restricted supply.

An interesting recent finding is that some binary mixtures of nonionic detergents show a markedly better efficiency than either of the components separately. A few commercial products based on this finding have been marketed, but they seem to be all difficult to biodegrade. Standard detergency theory and emulsion theory are difficult to apply to the wool scouring process, so detergent development has been largely empirical.

The following statements about the different types of detergent systems appear to be well established.

1. Neutral detergent scouring, i.e., using nonionic detergent alone with no builder, is suitable for many wools and has positive advantages with coarse wools that have a natural high alkalinity. There seems to be no advantage in using a neutral builder such as sodium sulfate in neutral systems. With fine wools, neutral scouring may lead to a product with a low pH, which is likely to give processing problems due to static electricity. The work of Henning [35] indicates an optimum pH of 9.5 for scoured wool.
2. Alkaline detergent scouring may give better results with some low yielding fine wools that are more difficult to scour neutral. In this case, alkali would be added to the first scouring bowl only.
3. Wool grease recovery is higher from neutral scours, but the quality of the grease is lower. Other factors such as the concentration of water-soluble substances in the scouring liquor also affect wool grease recovery, and the effect of pH may be minor.
4. Color, loftiness, strength, and brittleness of the wool are not affected by the detergent system if the scouring is well controlled. However, the risk of wool damage from poor control is greater for alkaline systems in general, and is particularly large in soap scouring.

Anionic Detergents

It was realized at an early date that anionic detergents will exhaust onto wool from a scouring bowl, and for this reason they were considered for many years to be unsuitable for wool scouring. Recently, because of the development of different types of anionics and their relatively low cost compared with other detergents, some mills, particularly in Japan, have started using mixtures of anionic and nonionic detergents for scouring. The anionic detergent exhausts partially onto the wool and is reputed to reduce the incidence of static electricity during processing, with beneficial effects especially in spinning.

It is believed that these mixtures contain 30-50% anionics, but there are no references in the literature either to detergent performance or to the processing characteristics of the scoured wool.

Synergistic effects, as mentioned above for binary mixtures of nonionic detergents, may possibly improve the efficiency of some anionic-nonionic mixtures.

Detergent Utilization

The available evidence suggests that detergent consumption is the least variable of all scouring parameters. A given type of wool will require a certain amount of detergent for adequate scouring, and this amount does not seem to be affected by operating procedures within the normal range of variation. The number of bowls, the countercurrent flow rate, and the pattern of detergent distribution between the different bowls seem to have no detectable effect on detergent consumption.

It is known that higher temperatures give a better detergent efficiency [36] up to 80°C, but practical temperatures are limited to fairly narrow ranges for ordinary scouring machines. Soap scours have an upper limit of 55°C on account of the danger of alkali damage to the wool; nonionic detergent scours lose efficiency rather rapidly below 60°C; and heat losses and machine tending become troublesome over 65°C; so in practice soap scours always operate in the range of 45-55°C and detergent scours in the range of 55-65°C.

Recently claims have been made [37] that detergent consumption is drastically reduced in scouring machines running under the Lo-flo system (see Sec. 4.3). It is speculated that this reduction is due to an improvement in efficiency stemming from the large number of bowls or a better utilization of the scouring ability of the suint. More recent unpublished results suggest that these claims may not be justified.

Although the pattern of detergent distribution between the bowls has no effect on detergent efficiency, it is in effect determined by foaming, and in practice little variation is possible. Thus, supposing the detergent requirement is 0.7 kg per 100 kg of greasy wool (a typical figure), it might be added to three bowls in the proportions 0.2 kg to bowl 1, 0.3 kg to bowl 2, and 0.2 kg to bowl 3. The same scouring result would be obtained by adding 0.2 kg to bowl 1 and 0.5 kg to bowl 3, with no additions to bowl 2; but additions at this rate would give unmanageable foam levels in bowl 3. However, it may be possible to add 0.4 kg to bowl 1 and 0.3 kg to bowl 2, since foaming is suppressed in liquors with higher grease contents.

Detergent is redistributed between the bowls by two mechanisms: first, detergent is carried forward in liquor on the wool; second, detergent is carried back through the machine by the countercurrent flow. As a result of these two mechanisms, all bowls will receive a proportion of the detergent, irrespective of where the additions are made. Under normal operating conditions, the net flow of detergent is backward, so a useful practical rule is to add the detergent as near the rinse end of the machine as is permitted by foaming considerations.

Detergent consumption in scouring varies considerably with the wool type. High yielding, coarse wools such as New Zealand fleece wools may require as little as 0.15% of nonionic detergent on the weight of greasy wool, whereas low yielding, fine wools may require as much as 1.5%. If soda ash is used as a builder, detergent quantities are reduced, but the total cost remains about the same. Corresponding quantities of soap range from 0.2-2%, together with 0.5-3% of soda ash.

2.5 Suint Scouring

Suint contains many soaplike and other surface-active substances, and has been known for a long time to possess good detergency under appropriate conditions [38, 39]. This is allowed for in normal scouring procedures (see Sec. 2.7), where common practice is to start the first scouring bowl with a detergent solution but not to make any further additions of detergent as scouring proceeds; the suint rapidly builds up to a level where the scouring efficiency of the liquor is equal to that of the original detergent liquor.

However, suint scouring in this section refers to that mode of scouring where the first bowl contains water at ambient temperature. Usually there is a continuous addition of water to the bowl and an overflow to drain. A bowl operated in this way removes a high proportion of the suint, a small proportion of the grease [40], and a moderate proportion of the dirt.

Advocates of suint scouring claim that it removes a significant proportion of the wool contaminants without cost (no detergent is used), that detergent consumption is thereby reduced, that the quality of the grease recovered is better because of the lower dirt content of the first hot scouring bowl, and that the color of the scoured wool is improved. Although some evidence has been published in support of these claims [41], there are also reasons for doubting them: the detergency contribution of the suint toward grease removal is lost by using a cold bowl; also, grease-dirt complexes (which might affect grease quality) are probably formed before scouring and are not appreciably removed by a suint bowl.

Even if small advantages are present, they must give way to the modern requirements of low water consumption and effective effluent treatment, both of which require a close approach to steady-state countercurrent operation (Sec. 1.4). Such an approach cannot be made if a suint bowl is used, and it seems probable that suint scouring will become increasingly rare in the future.

2.6 Transference of Contaminants from Wool to Liquor

A knowledge of the mechanisms by means of which the wool contaminants are detached from the fibers and transferred to the scouring liquor is of help in trying to forecast what effects will be produced when various operating parameters of the scouring process are changed. However, these mechanisms are still imperfectly understood, and the picture painted below should be regarded as provisional.

Suint

Included in this term are all the water solubles in the fleece, some of which derive from excreta. Suint itself is very readily soluble in water, and in fact is hygroscopic, so its removal in the scouring bowl is simply a matter of dissolution. Surprisingly, analysis of wool samples before and after the first scouring bowl shows that removal of water solubles is not as rapid as one would expect, and an appreciable quantity remains on the wool. This may be because (a) movement of liquor through clumps of fleece is hindered, (b) some water-soluble fleece components are not as soluble as pure suint, or (c) some of the suint is bound into grease-dirt-suint complexes that are slower to react with the scouring liquor. Since removal of the grease from the wool is more rapid than removal of water solubles [42] it appears that alternative (b) is most likely.

Grease

Unoxidized grease emulsifies in the classic manner, giving a droplet consisting of almost pure grease stabilized by a surface layer of detergent. Oxidized grease also forms a stable droplet, in this case a complex one containing dirt and occluded liquor subdroplets. Some oxidized grease probably sticks to mineral particles. Emulsification of both oxidized and unoxidized grease takes place very rapidly in the hot scouring liquor, and removal of grease from the wool in the first bowl is usually around 90%. (See also Sec. 3.1 and Fig. 5.11.)

Dirt

Some dirt finishes up inside oxidized grease droplets. Since very few of these droplets are more than 10 μm in diameter [43], the dirt particles must be even smaller. Some are clay particles from pasture soil, and the others are organic. Evidence on the nature of the organic particles is scanty, but many seem to be derived from excreta. It is highly probable that this fine dirt becomes incorporated into the grease before shearing and that the emulsification process takes place in the same way as for straight grease.

Coarse dirt particles may be sand grains or larger cell fragments. Again, many would become smeared with grease while still on the sheep, and when they are removed in the scour a certain proportion of grease adheres to them. Coarse particles may be stuck to the wool by suint or grease, or may simply be trapped between the fibers. When the adhesive materials have been dissolved or emulsified, the coarse dirt must be removed by mechanical action. Detergency mechanisms do not apply to particles of this size.

It is in the removal of these particles that nonfelting scouring machines have failed. In an immersion bowl, fibers tend to be mutually repulsive, so that staples expand as the fibers float apart. Dirt particles trapped between the fibers are released and are carried away by the liquor. When the wool staples are prevented from expanding, the wool mass acts as a filter and coarse particles cannot be removed. Besides the small particles associated with the oxidized grease, there are a number which still adhere to or perhaps have been redeposited on the fibers after the grease and suint have been substantially removed. They impart a gray color to the wool, and are usually removed by rinsing in the last bowl or two. Although they constitute only a fraction of a percent by weight of the scoured wool, they are considered important on account of their effect on color. Many scourers believe that they can be satisfactorily removed only by using very large quantities of rinse water.

Transfer of these particles from the wool to the liquor is undoubtedly controlled by detergency mechanisms. It is likely that they will be studied in some detail in the future, since their behavior may be important in many proposals for total water treatment and recycling.

2.7 Liquor Systems

As mentioned in Sec. 1.4, it is possible to specify an ideal scouring system that operates under steady-state conditions; in practice, conditions should be arranged to approach this ideal as closely as possible. In this section, the ideal system is discussed in some detail, and then practical systems are described in relation to the ideal.

The Ideal System

In Fig. 5.6 the wool passes through bowls 1 to 4, in that order. Detergent is added to one or more of the first three bowls, and water to the last. The water flows back from bowl 4 to bowl 1 via bowls 3 and 2 and is then discharged. Under steady-state conditions, the quantities of materials added to the machine at points A, B, and C exactly equal the quantities discharged at points D and E. It is easy to see that since only wet, clean wool leaves at point D, all of the contaminants removed from the wool, plus the chemicals added at B, must leave at E. It is also easy to appreciate that if the flow of water from C to E is increased, all the liquors will be cleaner and fewer bowls will be needed to achieve the same scouring result. Similarly, if the flow of water is reduced, the liquors will become dirtier and more bowls will be required.

If a device is installed that is capable of removing contaminants from the scouring liquors, the liquors will of course be cleaner. A

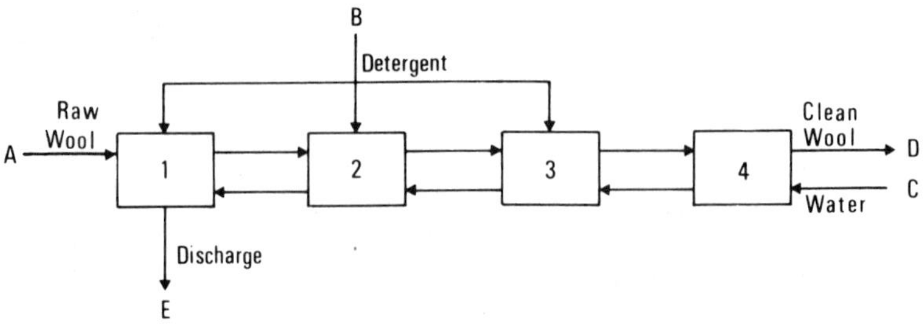

Figure 5.6 The ideal system of steady-state scouring. Only four bowls are shown, but there may be any number.

system of this sort is shown in Fig. 5.7, where the contaminant-removal device is attached to the first bowl only. (Even though contaminants are removed from the first bowl only, all the liquors will be cleaner since fewer contaminants will be carried forward to the other bowls in the liquor entrained in the wool.) Advantage can be taken of the contaminant removal device by reducing either the number of bowls or the countercurrent flow rate.

Considering the contaminant removal loop in Fig. 5.7, a higher efficiency in the removal device will lead to a lower contaminant concentration in bowl 1 liquor. A higher circulation rate through the removal device will also lead to a lower concentration in bowl 1 liquor (assuming that efficiency is not affected by circulation rate) [44]. Summarizing the important operating relationships in the ideal system:

1. The higher the number of bowls, the lower the countercurrent flow rate required.

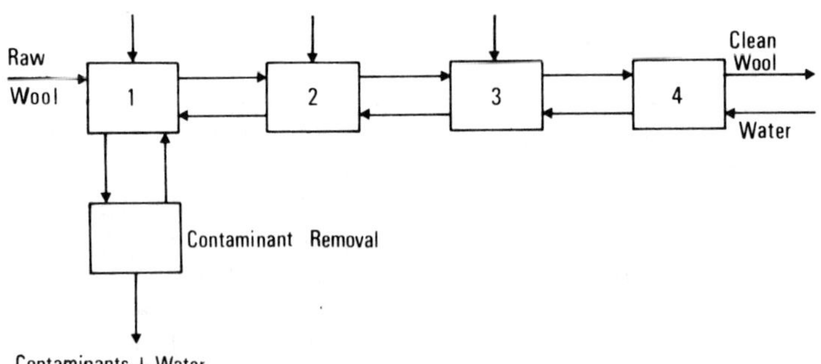

Figure 5.7 An ideal system with continuous contaminant removal from bowl 1.

2. The more efficient the contaminant removal device, the lower the countercurrent flow rate required (or the smaller the number of bowls required).
3. The greater the flow rate round the contaminant removal loop, the lower the countercurrent flow rate required (or the smaller the number of bowls required).

Limitations to the Ideal System

In practice, the approach to the ideal system described above is limited by a number of factors, which are discussed in the following subsections.

Number of bowls. Nobody today would install a machine with fewer than four bowls. Although good scouring is quite possible with one, two, or three bowls, detergent, water, and heat consumptions become uneconomical. Most new installations in recent years have called for five bowls, but scouring machines with four or six bowls are common. The introduction of very small bowls in recent times has led to the promotion of seven, eight, and higher numbers of bowls with the primary object of reducing water consumption [23]. It seem likely that six will become an accepted minimum.

However, for many years to come most wool will be scoured in conventional four-, five-, or six-bowl scouring machines. The minimum water consumption in these machines is about 2 liters per kg greasy wool for coarse wools and 4 liters per kg for fine wools in the scouring bowls, plus an additional 4-10 liters per kg in the rinse bowls.

Nearly all scouring machines would have three scouring bowls, i.e., bowls in which the bulk of the contaminants are removed. The remaining bowls—one in a four-bowl set, two in a five-bowl set, etc.—would be rinse bowls to which no detergent is added. The function of the rinse bowls is to remove a very small quantity of fine dirt (Sec. 1.1). There is little doubt that the removal of this dirt is assisted by the detergent carried forward on the wool from the earlier bowls. The primary restraint on the number of bowls used in practical installations is economic. A machine with too large a number has too high a capital cost, while one with too small a number has too high an operating cost.

Use of a suint bowl. In the ideal system the first bowl cannot be a suint bowl, since the purpose of a suint bowl is negated if it receives a countercurrent flow of hot detergent liquor from the second bowl. However, it is possible to add a suint bowl to the front of an ideal system, the operation of which remains otherwise unchanged. Water consumption would be increased and it is probable that overall cost would be higher. Many forms of effluent treatment would be prejudiced by the presence of a suint bowl.

Contaminant removal efficiency. Only one method—separation of suspended contaminants in a gravitational field—is currently used to purify liquors during scouring. (Air flotation has been used in the past; other methods, described in Sec. 4, are effluent treatment processes that take place after the liquor has been discharged from the machine.) The separation may take place under normal gravity (settling) or under artificial gravity (centrifuging). Another gravitational apparatus, the hydrocyclone, has been tried many times [45] but has not been found very useful.

Gravity settling is used to remove those particles that have a specific gravity appreciably greater than that of the liquor and thus sink readily to the bottom. Centrifuging is used to remove both particles that sink and light ones (grease) that float. Gravity settling has very low efficiency and is used primarily to remove coarse particles that would otherwise cause high erosion rates in various parts of the liquor system. The ordinary scouring bowl permits the settling of dirt particles through the false bottom. They accumulate as a sludge and in the so-called self-cleaning bowl are removed by a screw conveyor, together with a relatively large quantity of liquor. Gravity settling plays a special role in the Wool Research Organization of New Zealand (WRONZ) system [46].

Centrifuging, although much more efficient than settling, can still remove only a part of the suspended particles. The light phase normally contains no more than 40% of the grease washed from the wool, and large quantities of fine dirt always remain in the liquor after centrifuging.

Because of these limitations, the minimum countercurrent flow rates that can be used with a five-bowl machine in steady-state operation are believed to be between 2 and 6 liters per kg of greasy wool, depending on the wool type. Considerable doubt attaches to these figures, since in practice scouring machines are always washed out before steady-state conditions have been unequivocally attained. Few mills in practice have reached figures below 8 liters per kg, which has been regarded as a minimum usage in Japan for several years, and is specified in some European legislation with the implication that it represents the best available technology. It seems highly probable that additional rinse water will have to be used with lower yielding wool types to achieve satisfactory color in the scoured wool.

Recent work [23] has shown that by changing scouring techniques, the properties of the scouring liquor can be changed so that centrifuging efficiencies are greatly increased. The full practical implications of these developments have still to be worked out.

Flow rate in the contaminant removal loop. Commercial centrifuges suitable for use with wool-scouring liquors are available in only a small range of sizes. The traditional machines had a liquor throughput capacity of 4-5 m^3/hr, and more recently machines with a capacity of

8 m^3/hr have been used. A rough rule of thumb is that circulation rate round the centrifuge loop should be about 1 m^3 for every 100 kg of greasy wool, so a scouring machine with a throughput of 1000 kg/hr would require two 5 m^3/hr machines or a single 8 m^3/hr machine.

Centrifuges are so expensive that the extra contaminant removal achieved by increasing the circulation rate via extra centrifuging capacity is not ususally economical. Existing centrifuges should always be run at the maximum throughput rate.

Currently Used Systems

The conventional system. A diagram of the conventional system is given in Fig. 5.8. The first three bowls are scouring bowls and the last two are rinse bowls.

Detergent is added to the first three bowls, or if alkali is being used, it alone will be added to the first bowl and detergent to the next two. Rinse water is added to bowl 5, flows back to bowl 4, and is then discharged. Additional countercurrent flow takes place from bowl 4 to bowl 1. Liquor is pumped from bowl 1 through a three-phase centrifuge and back to the bowl. The sludge discharge from the centrifuge goes to drain, possibly with some additional liquor from the centrifuge. Periodic sludge discharges to drain are also made from the bottom of bowls 1 and 2. The total effluent comprises sludge, liquor, and rinse water.

If the wool throughput is 1000 kg/hr of a medium-yield fine wool, the feed rate to the centrifuge would be 10 m^3/hr, the cream rate would be about 150 kg/hr of an emulsion containing 30-40% grease, and the sludge discharge would be about 1.7 m^3/hr with a total solids content of 6-7%. If the total water consumption is 10 m^3/hr, the liquor discharge would be set at 3.5-4.0 m^3/hr, and the rinse water would be about 4.5 m^3/hr to make up the difference.

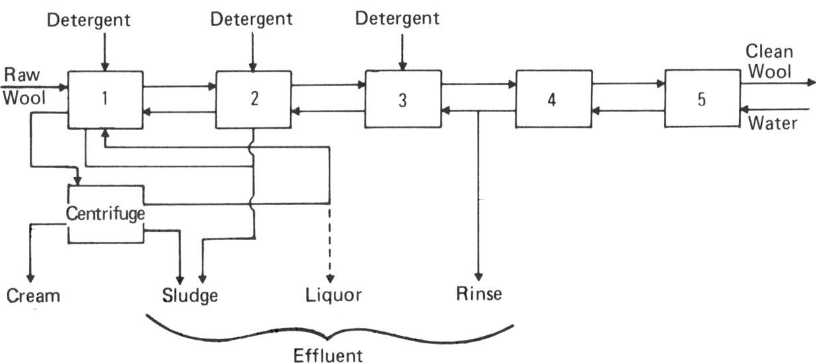

Figure 5.8 Conventional scouring using three scouring bowls and two rinse bowls.

Common variations on this arrangement are for rinse water to be discharged from bowl 5 instead of or as well as from bowl 4, and for bowl 2 liquor to be circulated through the centrifuge for short periods from time to time.

Conventional system with suint bowl. A diagram of this system is given in Fig. 5.9. The flow through the suint bowl would be typically about 4 liters per kg of greasy wool, but the corresponding reduction in the flow through the other bowls would probably not be more than 2 liters per kg, giving a net increase in water consumption over the conventional system.

The WRONZ system. This scouring system [46] includes much more than the liquor system, but only that aspect of it will be dealt with here. Its principal features are: (a) the use of modified scouring bowls (either special new bowls or modified existing bowls) that permit the controlled desludging of main bowl and side bowl; (b) full countercurrent flow through the machine, with discharge from one point only; and (c) the use of a large gravity settling tank in conjunction with the first bowl.

An important feature which this system has in common with the pioneering CSIRO jet scouring machines of the 1960s is that all liquor circulation is from the *bottoms* of the bowls or side bowls, thereby delaying the accumulation of sludge. Problems associated with abrasion caused by circulating bowlbottom liquors are avoided by introducing a heavy solids tank (HST), in which the liquors are settled before passing to the grease centrifuge.

Referring to Fig. 5.10, bowl 1 liquor is fed to the HST, from which settled liquor passes to the grease centrifuge. Occasionally,

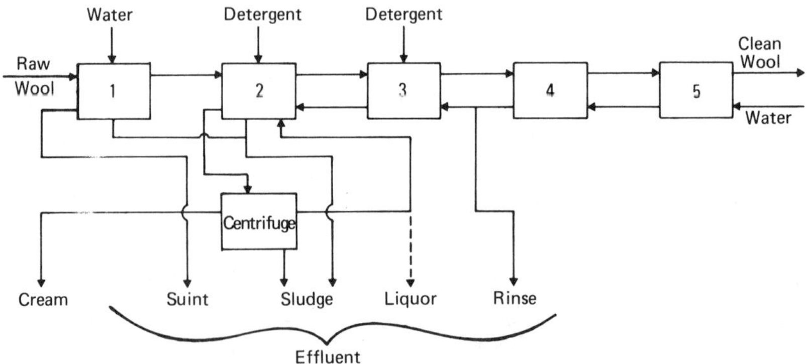

Figure 5.9 Scouring with a suint bowl. Water consumption is greater than in conventional scouring.

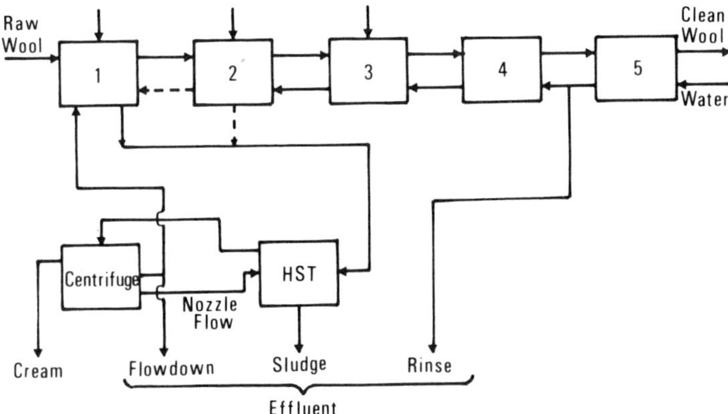

Figure 5.10 The WRONZ system. Water consumption is reduced by rationalized liquor handling.

bowl 1 liquor is replaced by bowl 2 liquor as the HST feed. Since the return from the centrifuge is always to bowl 1, the total feed from bowl 2 to the HST never exceeds the required countercurrent flow from bowl 2 to bowl 1, which it replaces either partly or fully.

From the grease centrifuge, the sludge discharge is returned to the HST and the purified liquor returns to the bowl. Part of this liquor, called the *flowdown* is discharged to drain. The sludge from the HST is discharged periodically, its total solids content being in the range of 15-25%. The total effluent comprises flowdown, sludge, and rinse water.

The flowdown in this process varies from 1.5-3.5 liters per kg, depending on the wool type. In all existing installations the main bowls are washed out at intervals ranging from a few days to a month.

The Lo-flo system. This system was designed for effluent treatment rather than scouring improvement and is described in Sec. 4.

2.8 Quality Control and Product Testing

Quality Control

Scouring. The controllable variables in scouring are wool feed rate, detergent addition rate, bowl temperatures, and water input. Temperatures are never changed, but all the other variables are used in different ways to make sure that the quality of the scoured wool remains satisfactory.

Wool feed rate is usually predetermined for a given wool type. Lower yielding and shorter types are run at lower feed rates than the others, but feed rate is usually not varied during a run. Water throughput through the rinse bowls may be increased if it is considered that the wool needs brightening. These are machine adjustments that help to maintain quality, but they are not used for feedback control. The only real feedback control is of the detergent additions.

The feedback parameters are the residual grease content (r.g.) of the wool and the subjective assessment of quality—color, brightness, felting, loftiness, etc. Virtually every mill makes continuous measurements of r.g. and adjust detergent additions accordingly. However, there is no clear correlation between r.g. and detergent addition rate, so automatic control is impossible. Manual adjustments are made on the basis of the history of the run to that point, the subjective quality of the scoured wool, and the operator's experience. This is where most of the scourer's art is exercised.

Accurate measurement of r.g. is a long procedure, soxhlet extraction being the universally recognized method. Dichloromethane (DCM, methylene chloride) is the most commonly used solvent, though others are sometimes used for special purposes. Since a test that will not give a result for several hours is of no use for feedback control, many attempts have been made to devise an acceptably accurate method that can be carried out quickly.

The WIRA rapid method [47] achieved widespread use after its introduction in 1950. However, it contains a systematic error which, while fairly small and constant in the hands of a skilled operator, tends to be large and variable in the hands of the average mill worker. Various improvements to or variations of the WIRA method have been put forward from time to time; with the introduction of the automatic balance, apparatus is now available which can give a result equivalent in accuracy to the soxhlet method in about 15 min and can be used by relatively unskilled persons.

With conventional scouring bowls the large liquor volume prevents rapid changes in liquor properties, and an r.g. test every half hour is quite adequate for process control. With modern small bowls, control is more difficult.

Another method used on account of its simplicity is the *turbidity* method. In this method, a weighed sample of wool is shaken up with a fixed quantity of a standard detergent solution. The turbidity of the solution, presumed due to emulsified grease, is then compared to solutions of standard turbidity representing known r.g.'s, e.g., 0.4%, 0.8%, and 1.2%. Despite the large uncertainties of the method, it is quite widely used with apparent success. This success may be a reflection of the skill of the operator and the weight given to subjective assessment of the scoured wool, rather than of the value of the method.

Drying. The traditional way of handling wool from the end of the scouring machine is to pass it through the dryer and convey it, usually pneumatically, to a large storage bin, where it is left long enough for the moisture to spread evenly through the bulk. The wool is then either packed or conveyed to the card room.

This practice is based on the acceptance that wool cannot be dried evenly. Frequent tests are made of wool regain (moisture content calculated as a percentage of the dry fiber weight), and dryer settings are adjusted manually to try to achieve the desired regain in the equilibrated wool mass in the storage bin. Dryer temperature, speed (residence time), and air exhaust rate can all be adjusted on drum dryers; but in practice the first is most commonly used, the second less frequently, and the third hardly at all.

The adoption of wool feed rate controllers [16] makes uniform drying possible, and continuous feedback control of drying becomes practicable [48]. Adequate sensors for wool regain are not available, and control is through a combination of temperature and humidity sensors, the latter controlling the air exhaust rate.

For internal mill purposes, wool regain may be measured either by drying and weighing or by electronic means. Instruments based on the latter principle are unreliable in use; to give consistent readings the wool sample must be prepared with extreme care. Most mills therefore prefer the CSIRO direct-reading regain tester [49], which incorporates the drying, weighing, and calculating procedures into an unskilled operation.

Product Testing

The important properties of scoured wool are residual grease content, regain, and color. For combing wools, felting is also important. If the wool is to be used in the same mill, testing is usually confined to r.g. and regain, but when the wool is for sale, subjective assessments are also very important. These include brightness, hand, bulkiness, and comparison with a commercial sample.

Color measurement is possible [50] but not usually practiced in industry because the instruments are very expensive. Regain measurements are vitally important in commercial transactions and are usually carried out by independent wool-testing laboratories. Residual grease content of wool for sale is usually specified as the dichloromethane extract, but some buyers still specify an ethanol extract. (For soap scoured wool, the difference between a diethylether extract and an ethanol extract used to be regarded as the soap content of the wool.)

3. WOOL GREASE RECOVERY

3.1 The Stability of Wool Grease Emulsions

When contaminants are removed from the wool, they pass into suspension in the scouring liquor (see Sec. 2.6) and form suspended particles

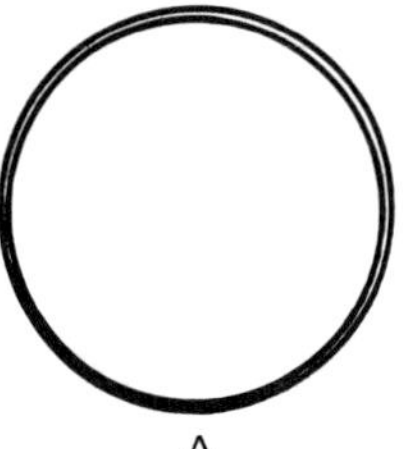

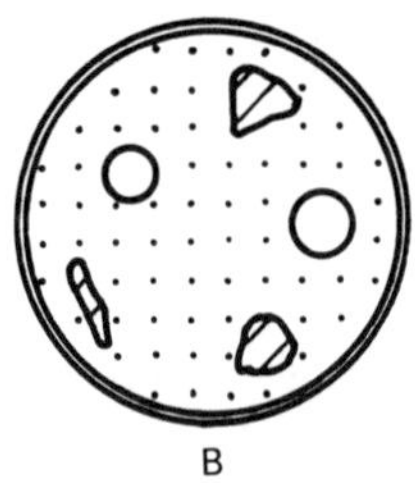

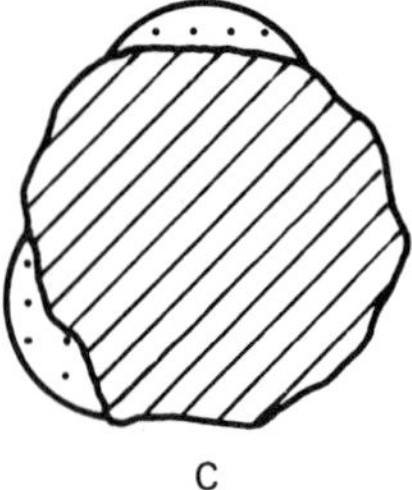

Figure 5.11 Different types of suspended particles in scouring liquor. (A) Unoxidized grease droplet with surface skin; (B) oxidized grease droplet containing dirt particles and occluded liquor drops; (C) mineral particle with adhering oxidized grease.

of three sorts (Fig. 5.11). Unoxidized grease forms classic emulsion droplets consisting of a sphere of melted grease with a surface skin of detergent molecules. Oxidized grease (see Sec. 1.1) forms complex emulsion droplets that contain dirt, occluded scouring liquor, and other components. Most of these emulsion droplets are between 1 and 10 μm in diameter, so the dirt particles they contain must have a maximum diameter of a few microns. Other dirt particles, some with oxidized grease sticking to them, are simply suspended in the liquor.

If the liquor is settled without cooling, most of the dirt particles and some of the oxidized grease droplets sink to the bottom to form a sludge. The sludge undergoes some compaction with time, but rarely achieves a solids content higher than 15%. Some of the unoxidized grease droplets float to the surface, but most of the emulsified grease, whether unoxidized or oxidized, remains suspended in the liquor.

If the hot liquor is centrifuged, better separation occurs. Virtually all of the unoxidized grease droplets float to the surface to form a cream, but they do not coalesce into a grease layer. Most of the dirt particles, and a proportion of the oxidized grease droplets, sink to the bottom, and the sludge they form may have a solids content as high as 35%. Of the solids in the sludge, about 20% may be grease. Up to 50% of the total grease may remain emulsified in the liquor, and this may hold a substantial proportion of the total dirt.

The liquor remaining after the cream and sludge have been removed from a centrifuged liquor is an emulsion of oxidized grease droplets (containing many other contaminants) in a solution of detergent, builder, suint, and other water-soluble substances from the fleece. The emulsion may be broken (flocculated, destabilized) in a number of ways, including reduction in pH and addition of polyvalent cations. An effect equivalent to the addition of cations is produced when water input to the machine is reduced, thus causing an increase in the concentrations of suint, detergent, builder, and other dissolved solids. The destabilization is manifested as an increase in the quantity of wool grease recovered in the centrifuge; a relation between recovery and dissolved solids concentration is shown in Fig. 5.12.

The extra recovery of grease is due to a combination of two effects. The presence of cations causes flocculation, but the flocculated particles may sink or float according to the density of the liquor. As the dissolved solids content of the liquor increases, so does its density, and more and more of the flocculated particles float. In the extreme case, grease recovery may increase from 30-40%—for many years considered to be the maximum possible [51]—to 60-90%. (Grease recoveries are expressed as a percentage of the grease on the raw wool.) When recovery is very high, virtually no grease remains in the middle layer, the unrecovered grease being all in the sludge.

Wool grease emulsions may also be broken by adding adsorptive materials such as bentonite to them. The grease droplets stick to the adsorbants, which may be removed by settling or other means. One variation of this method that has been used to recover wool grease is aeration (flotation). Grease droplets stick to the surfaces of air bubbles, which float to the surface of the liquor to form a greasy froth. This may be removed and grease recovered from it.

3.2 Recovery Systems

Historically, several different methods have been used to recover wool grease. Before the sludge-discharge centrifuge was invented, there was no way of recovering grease continuously and recycling the liquor back to the scouring bowl. Methods were therefore designed for use with the liquor after it had been discharged from the scouring machine.

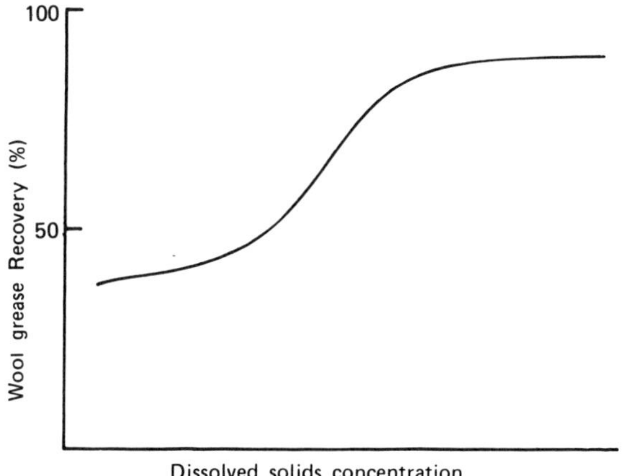

Figure 5.12 Destabilization of a wool scouring liquor by increased dissolved solids concentration. The greater the destabilization, the higher the wool grease recovery.

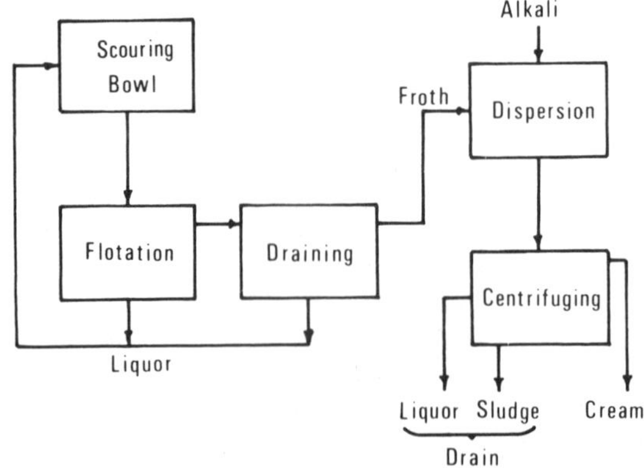

Figure 5.13 Recovery of wool grease by flotation.

They included acid cracking, solvent extraction, and a crude form of flotation [4]. Some of these have survived as effluent treatments and are described under that heading.

Modern scouring demands continuous removal of wool grease from the liquor by a method that does not affect its scouring ability. Only flotation and centrifuging are suitable. The equipment used for flotation is a modification of the flotation cell developed for ore enrichment. After the liquor has been aerated in the cell, the foam is removed, drained, dispersed in an alkali solution, and centrifuged (Fig. 5.13). Recovery of grease in the foam is higher than could be obtained by straight centrifuging of the liquor, but all the additional grease is lost when the dispersed foam is centrifuged. It is probable that with today's knowledge of the properties of wool grease emulsions, all the grease in the foam could be recovered. However, other less cumbersome methods for increasing grease recovery are now available.

For various reasons, flotation dropped out of use after a brief vogue in the 1950s [52]. It was still used in the USSR until quite recently [53], but today centrifuging is the only method used for recovering wool grease from emulsion scouring liquors.

3.3 Centrifuging

General Considerations

The effective centrifuging of wool scouring liquors was made possible by the invention of the sludge-discharge centrifuge and its subsequent commercialization by the De Laval (now Alfa Laval) company [4]. The machine is a vertical centrifuge (one that spins about a vertical axis) in which sludge is discharged from nozzles set in the periphery of the bowl.

The centrifuge is always connected to the first scouring bowl, and liquor is pumped from the bowl to the centrifuge and back continuously while the scouring machine is operating. In some modern centrifuges, sludge is discharged intermittently by opening the bowl (i.e., separating the top and bottom parts hydraulically and then closing them again) instead of through nozzles. The three liquor phases—cream, middle layer, and sludge—are all discharged separately by the centrifuge. Usually the sludge is removed from the system, the middle liquor is returned to the scouring bowl, and the cream is treated to separate the grease. The separation of phases under gravity is represented diagrammatically in Fig. 5.14.

There are no definite interfaces between the cream and the middle layer, or between the middle layer and the sludge. Rather, the concentration of grease diminishes gradually from the surface downward, while the concentration of dirt increases. The levels A and B represent the points at which the liquor is separated in the centrifuge. Everything above A is the cream, and everything below B is the sludge. Clearly, if the separation level A is raised, the cream will be less in quantity and higher in grease concentration. Lowering level A will give a larger volume of thinner cream. Similarly, raising or lowering level B will give, respectively, more thin sludge or less thick sludge. In the centrifuge, changes in level A are made by changing the size of the *gravity disk*, a variable dam over which the middle layer flows. The larger the gravity disk, the smaller the volume of cream. Changes in level B are made by changing the size of the sludge discharge nozzles. Larger nozzles give a larger flow of sludge, with an effective raising of level B. Similarly, a higher feed rate of liquor to the centrifuge effectively raises the level of the liquor surface to give an effect equivalent to the lowering of level A.

Changes in nozzle flow and feed rate also have the effect of changing the residence time of the liquor in the centrifuge. A reduction

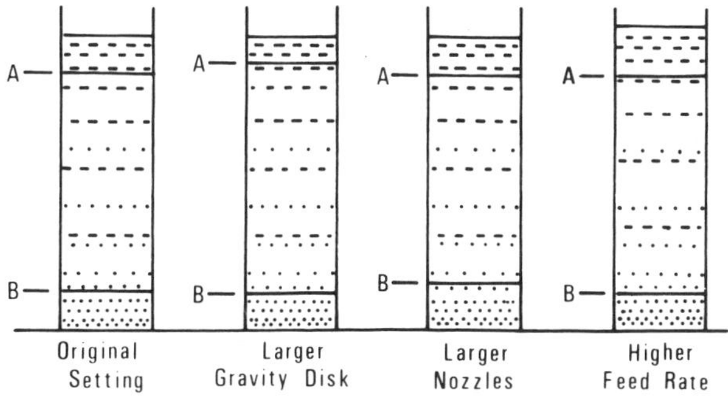

Figure 5.14 Diagramatic representation of the effects of various machine adjustments on the phase separations in a three-phase nozzle discharge centrifuge.

in residence time reduces the concentration gradient for both grease and dirt, and gives a thinner cream, a thinner sludge, and a middle layer with more dirt and grease. In practice, nozzle sizes are not used to control the operation of the machine. Small nozzles give a smaller volume of thicker sludge, which is preferable on general grounds, but they also block more readily. Nozzles become steadily larger through wear, and have to be replaced when the sludge volume becomes too high.

Usually, cream production from the centrfuge is controlled by setting the gravity disk size and the feed rate. First, a gravity disk is selected to give a cream of medium consistency. The feed rate is then used as fine control to thicken or thin the cream as required. The operation of opening bowl machines is essentially similar to nozzle discharge machines. Since sludge discharge is intermittent, the cream consistency tends to vary from thick to thin between bowl openings, but this has not been found to cause any problems.

Practical Considerations

It is clear from Fig. 5.14 that the lower the level A, the more grease will be removed in the cream. For maximum recovery, therefore, it would be expected that a fairly thin cream should be produced. This is in line with the requirement that the liquor circulation rate through the centrifuge should be as high as possible (Sec. 2.7). However, since each unit of liquor passes on the average several times through the centrifuge, the single-pass recovery is not of primary importance; grease not captured on one pass is likely to be recovered on a later pass. Also, we know that grease in the deeper layers is predominantly oxidized grease that will not separate under gravity (Sec. 3.1). There is little point in skimming these deep layers off with the cream, since the oxidized grease will eventually be lost in the secondary centrifuge.

Another argument against producing a thin cream is that it would require two more centrifugings to give anhydrous wool grease, whereas if a thicker cream is produced, a single extra centrifuging will suffice. Nearly all new installations today are two-stage (rather than three-stage) plants, in which a cream containing about 40% grease is produced in the first stage. The recovery of grease from the cream is described in the next section.

A fairly widespread practice is to heat the liquor before centrifuging. Some workers have claimed that the temperature of centrifuging has no effect on grease recovery [54], whereas others obviously believe that recovery is better at higher temperatures. It is possible that temperature effects are linked to other liquor variables such as pH or dissolved solids concentration, so may be present in some circumstances but not in others. For a scouring machine operating in an efficient manner with a countercurrent flow of perhaps 5-6 liters per kg (Fig. 5.8), grease recovery will be in the range 30-50%, depending on wool quality. When countercurrent rates fall to 2-4 liters per kg, recovery is likely to be 40-60%, and at 0.5-2.0 liters per kg recoveries will be 50-85%. The wool grease quality is always lower at higher recoveries, owing to the inclusion of highly contaminated oxidized grease droplets, and special refining procedures need to be used.

3.4 Wool Grease Refining

This section covers the extraction of anhydrous wool grease from the centrifuge cream and a few simple refining procedures that could be used in a scouring mill. It does not include any of the complicated refining, chemical modification, and fractionation processes normally carried out by wool grease refiners [55-57]. Many mills have tried refining their own grease to, say, USP or BP specifications. Very few have found it worthwhile, and the best strategy for disposal of wool grease by the mills seems to be to clean it by simple procedures and sell it to the refiners.

Treatment of the Primary-Centrifuge Cream

Cream from the primary centrifuge is usually delivered to a storage tank, where it is heated to 95-100°C. From the storage tank it is fed to a clarifier centrifuge, together with a quantity of boiling water. The water feed may be two or three times the cream volume. Mixing of the cream and water takes place in the entrance zone to the centrifuge, and then separation takes place to give a grease product with a water content of 0.5-2.0%. The grease is sometimes run over a hot plate as it leaves the centrifuge to further reduce the water content.

During this process, the continuous phase of the cream is diluted with the hot water, thus reducing both the density of the continuous phase and the concentration of surface-active materials. When the diluted cream separates, the oxidized grease droplets and the contaminants associated with them remain in the aqueous phase. The clean unoxidized grease droplets float to the surface, coalesce, and emerge as almost anhydrous grease.

There is a considerable improvement in the grease quality due to the secondary centrifuging, and losses of inferior grease may amount to 10% of the grease in the cream. Further quality improvement is obtained by a further washing and centrifuging, but in a three-stage process it is usual to produce a 70% cream instead of anhydrous grease from the secondary machine, and then to recover anhydrous grease from the tertiary machine. The secondary machine would be a concentrator and the third machine a clarifier, the concentrator being set to minimize losses of grease with the water, and the clarifier to prevent water getting into the grease. Grease losses in the tertiary machine are usually only 2-3%. The reader will have noted that the quantity of anhydrous grease is about 90% of the grease in the primary cream, so the overall recovery of grease will be 10% lower than the figures quoted in Sec. 3.3.

From the foregoing description, it will be appreciated that recentrifuging the primary cream is essentially a fractionation process that separates unoxidized from oxidized grease droplets. Clearly, a process of this nature is of little use with creams that are produced in very high yield. The extra grease in a high-recovery process is oxidized grease, and would simply be lost to drain if attempts were made to recover it by recentrifuging.

The problem is of too recent an origin to have an established solution, but two appraches have been described [23]. In one, known as *thermal cracking*, the cream is stored for at least 24 hr in a tank at 90°C. Grease with a fairly low moisture content separates at the surface, and dirty water with a low grease content settles at the bottom. Sometimes the grease is recentrifuged to reduce the dirt and moisture content; but if water is added, grease losses in the water are high, and if water is not added, dirt content of the grease remains high. Separation in this process is not clear, and there is often a middle layer of dirty emulsion. Also, the thermal exposure tends to darken the grease and so reduce its value.

In the second process, the cream is thoroughly mixed with water containing acid and a high ethylene oxide nonionic surfactant, and is then centrifuged hot. The dirt transfers to the aqueous layer with very little grease, and there is almost no interfacial material.

It seems certain that advances will be made in the very near future in the handling of high-yield creams. The ideal process would probably be one that recovered the oxidized and unoxidized grease separately.

Upgrading Recovered Wool Grease

The price paid by refiners for wool grease depends on color (by far the most important), acid value, and water content. A premium may be paid for detergent-free greases, i.e., those recovered from soap-soda scours. Improvements can be made in some of these properties by fairly simple processes, and mills are led to consider from time to time whether they could improve their grease income by bleaching and neutralizing. The fact that most of them do not do so is an indication of where the balance of profit normally lies. It is easy for a refiner to tell if a grease has been bleached or neutralized. These greases are less useful than the crude product, and the premium the the refiner is prepared to pay is less than the producer might expect. Therefore, unless a mill has a secondary customer (one who would normally buy from a refiner), it is usually not worthwhile to treat the crude grease. Even if secondary customers exist, there are often difficulties in maintaining appropriate quality standards, since the mill has no choice in the grade of grease produced, which is determined by the wool quality. However, for those who wish to use the processes, brief descriptions are given here.

Bleaching. Bleaching usually leads to an increase in acid value, so it precedes neutralizing if both processes are used. Hydrogen peroxide is probably used exclusively. Other oxidizing agents such as sodium chlorite are quite effective [58], but chemical residues are more difficult to remove from the grease. The quantity of peroxide used varies from about 0.2-1% of the grease weight. Sometimes small quantities of acid are added first. The wool grease is heated to 60-70°C in a stainless steel vessel, acid added if required, and then the peroxide added gradually. The reaction is strongly exothermic, and much foaming is caused by the gas evolved. Vigorous stirring is maintained, and this helps to control foaming. When foaming subsides, the temperature is raised to 95-100°C and maintained until all foam has disappeared. Water washing sometimes follows bleaching.

Neutralizing. It is difficult to reduce acid value below 0.5 by straightforward chemical neutralization. Special techniques are used by refiners to produce very low values. In the ordinary method suitable for mill use, the acid value of the crude grease must first be determined. The quantity of alkali (usually sodium carbonate) required for neutralization is then calculated stoichiometrically, and the quantity used may be 10% greater than this. The grease is melted in an appropriate vessel over a layer of water, often containing 10% of ethanol to prevent emulsification. The alkali is added to the aqueous layer, and very gentle stirring is continued until the reaction is complete. The two layers are then allowed to separate, and the top neutralized grease layer is run off. The whole operation is carried out at about 70°C. In wool grease refineries, the bottom layer is treated for alcohol and fatty acid recovery, but in a mill these procedures are usually impracticable.

4. WASTEWATER DISPOSAL

4.1 General Considerations

By assuming a greasy wool composition, a water usage, a grease recovery efficiency, and a crude solids settling efficiency, it is possible to work out the composition of the effluent. A calculation of this sort is shown in Tables 5.2 and 5.3 for a merino fleece wool yielding 63% scoured wool.

If the rinse and countercurrent discharges are assumed to be 5 liters per kg, each, the rinse water might contain 3% of the

Table 5.2 Distribution of Components in Wool Scouring Streams[a]

Component	Present in raw wool[b]	Added in scouring liquor[c]	Removed in cream	Removed in settling pit	Discharged with effluent
Grease	14		5.6	0.3	8.1
Suint	5		0.1	0.3	4.6
Dirt	15			3	12
Fiber[d]	55				
Water	11	1000	8.2	100	903
Detergent		1	0.1		0.9
Builder		1.5		0.1	1.4

[a]All figures are percentages of raw wool weight.
[b]Typical Australian merino wool.
[c]Total water usage 10 liters per kg.
[d]Includes vegetable matter.

chemical oxygen demand (COD) and the suspended solids (SS). The totals from Table 5.3 would then become: for the countercurrent discharge, COD 79,200, SS 37,000, and solvent extractables 20,000; for the rinse water, COD 2,400 and SS 1,100.

Whatever the arrangement, the discharge represents a massive pollution problem. A scouring machine processing 1000 kg/hr of greasy wool on three shifts, on the basis of the figures in Table 5.3, would discharge 220 m^3/day. Requirements for the discharge of effluents vary widely. For discharge to the environment (i.e., into rivers, lakes, or the sea), the required biological oxygen demand (BOD) may be 10-20 mg/liter, equivalent to a COD of 50-100, and the required SS 30-50. These figures represent virtually complete purification of the wastewater.

For discharge to a sewer, the upper limit for COD may be in the range of 300-1500, SS from 200-1000, and solvent extractables from 0-300. For some sewer discharges no limits are applied, but pollutants must be paid for according to the weight discharged.

Wool scouring wastes are particularly difficult to purify because of their strength and complex composition. Some of the impurities are dissolved, some are emulsified, and some are suspended (Sec. 3.1). Successful treatments have to remove all these classes of impurities, and often will include two steps, one for removing suspended and emulsified particles, and a subsequent one for dissolved impurities.

The first step, destabilization, has already been discussed in Sec. 3.1.

Simply removing the impurities from the water is not enough—they have to be removed in a disposable form. Thus, the solids should be in the form of a solid cake rather than a liquid sludge; the grease should preferably be recovered separately for sale in the wool grease market; the dirt cake should not be prone to odor producing decomposition.

The cost of effluent treatment may be significantly affected by scouring procedures. Thus, if grease recovery is high, there will be less grease in the effluent; dirt removed in wool opening reduces the pollution load; reduced water consumption gives a lower effluent volume; keeping rinse water and countercurrent discharge separate may simplify effluent treatment.

Accordingly, it makes sense for a mill with an effluent treatment problem to rationalize the scouring procedure before installing treatment plant. The Wool Research Organization of New Zealand (WRONZ) has been particularly successful in promoting rationalized scouring. The WRONZ system of scouring (Sec. 2.7) enables most of the dirt to be discharged separately as a fairly concentrated sludge, grease recovery to be increased, and wastewater volume to be reduced. Effluent treatment requirements following a WRONZ scour are sludge dewatering, treatment of a small volume of strong liquor, and perhaps treating rinse water.

Table 5.3 Effluent Strength (mg/liter) Calculated from Table 5.2

Component	Quantity[a]	COD[b]	SS[c]	Solvent extractables[c]
Grease	8.1	25,970	6,000	8,700
Suint	4.6	3,900		300
Dirt	12	8,090	13,000	
Water	903			
Detergent	0.9	2,890		1,000
Builder	1.4			
Total		40,850	19,000	10,000

[a]From Table 5.2 as percent of raw wool weight.
[b]Ref. 59.
[c]Estimated.

The use of a suint bowl adds significantly to the problem of effluent treatment (Fig. 5.9). It provides one more discharge containing large quantities of soluble and insoluble impurities which cannot be treated in the same centrifuge circuit as the hot scouring liquors. Either it must be discharged without inline treatment, or separate treatment facilities must be installed. Where wastewater treatment is a serious problem, it is doubtful whether the retention of a suint bowl could be justified.

At present, the development of processes for purifying scouring wastewater is proceeding very rapidly, but very few have been used sufficiently widely over a sufficient number of years to be regarded as established processes. In the descriptions of individual processes that follow later, they are dealt with more or less in the order of the extent to which they have been used.

4.2 Types of Processes

Biological Processes

Aerobic processes. The wastewater is oxygenated to encourage the growth of microorganisms, which feed on organic impurities and convert them to carbon dioxide, water, and oxidized compounds of nitrogen, sulfur, phosphorus, and other elements. Eventually, the aerated liquor will contain the dead remains of the microorganisms and the inorganic constituents that cannot be used as food. The insoluble constituents can be settled out as a sludge, leaving a clear water for discharge.

The advantages of biological oxidation are that it is cheap and it is capable of giving complete purification. Its disadvantages are that the microorganisms are susceptible to environmental changes (pH, temperature, etc) and poisons, and do not respond well to shock loads; some pollutants are not biodegradable; the process may produce a large volume of weak sludge (less than 5% solids content) which is very difficult to dewater; and it does not permit separate recovery of the grease.

Aeration processes may be slow or fast. The former require large lagoons with residence times up to several weeks, usually with floating aerators or other artificial oxygenators. The fast ones include the *activated sludge* process, where degradation rate is increased by recycling microorganisms, and high-rate trickling filters, where the microorganisms are grown on a stationary medium over which the liquor trickles.

Some biological processes use a combination of anaerobic and aerobic treatments, and some are combined with chemical treatments.

Anaerobic processes. These operate best at higher temperatures than aerobic processes, 40°C being a common operating temperature. The main end products are methane and carbon dioxide in a mixture *biogas* that can be used as a fuel. Anaerobic digestion has not found a use for treating scouring liquors—one large installation in Japan was discontinued after a few years. However, sludges from wool scouring have been anaerobically digested in municipal plants along with domestic sewage sludges.

Chemical Processes

In these processes, the wastewater is flocculated by adding a chemical, and the precipitated impurities are removed by settling, centrifuging, filtering, or air flotation. The advantages of the processes are that they can easily be controlled by feedback systems; their space requirements are small; and they sometimes produce a sludge that is easy to dewater. The disadvantages are that the chemicals used may be expensive; they do not remove dissolved impurities; they usually produce a very large sludge volume; and they do not give a separate recovery of wool grease.

Chemical processes have not been widely used for treating wool scouring wastes. Straight flocculation with polyvalent cations or polymeric flocculants gives very large sludge volumes and is too expensive. The only treatment that looks at all commercial in 1980 is the hot sulfuric acid process developed by McCracken [60].

Wet oxidation treatments such as the Zimpro process [61] are chemical degradative processes rather than flocculations. They are regarded as too expensive and have never been tried with wool scouring wastes.

Physical Separations

Methods such as filtration and centrifuging are of course quite ineffective in removing pollutants from scouring wastes. However, more powerful methods such as membrane processes (ultrafiltration, microfiltration, reverse osmosis) and evaporation can be very effective.

The membrane processes [62-65] do not usually separate more than a minor proportion of the dissolved impurities, and they can be very expensive if the membrane life is short. What is more, the sludge they produce is usually no thicker than that produced by straight settling in the WRONZ scouring system, and it must be dewatered by an extra treatment. These processes do not improve grease recovery.

Evaporation is an ideal process in the results it achieves. The impurities, both soluble and insoluble, are completely separated from the water, which can be recycled. Technical difficulties tend to make the process expensive, but it has been widely used in Japan for wool scouring liquors [66] and its use is spreading. In one version, additional wool grease is recovered by centrifuging the concentrated liquor from the evaporators.

Dissolved air flotation is a physical method of separating suspended solids. Like centrifuging and filtration, it is ineffective on straight scouring wastes but may be used with flocculated liquors. Tiny bubbles of air, previously dissolved under pressure, are formed when the pressure is released and carry the suspended particles to the surface.

4.3 Processes Used in Industry

Large-Area Biological Treatments

These are probably the most widely used of all processes. They are comparatively cheap to install and operate, give complete purification, and do not generate any problems in sludge disposal. Unfortunately, they need large areas of land, so are not applicable to mills in urban areas.

A common arrangement is first to treat the waste liquors in an anaerobic lagoon with a retention time of a few weeks. Usually a layer of grease forms on the surface and seals the lagoon from the air. The liquors then pass to an aerobic lagoon with a retention time of 1-2 weeks. Mechanical aerators keep the liquor turning over. Finally, the liquors pass to a very large, shallow lagoon with a retention time up to several months. The outflow from this lagoon may be of sufficient purity to discharge to a river, or may be further treated by spraying on to pasture. If spray irrigation is used, the minimum area required for a mill scouring 6 Mkg of greasy wool per year is about 20 hectares (50 acres) if annual rainfall does not exceed 750 mm.

Evaporation

In Japan a number of mills have been treating their effluents by evaporation for almost a decade [66]. The scouring units involved are all operated in a similar fashion and use about 8 liters of water per kg of raw wool. The effluent is concentrated by evaporation from about 5% total solids content to about 15%, and this concentrate is then fed directly to an incinerator, which reduces it to ash.

The most often used evaporators in Japan are plate heat exchangers. It is to be noted that the final concentration is quite low, and in fact is no higher than can be achieved by rationalized scouring using low water input. For evaporation down to a solids content of 80-85%, which would be the aim if the sludge is to be disposed of by landfill, plate heat exchangers would not be satisfactory.

Outside Japan, two mills in Europe have installed evaporation plants. One is a six-stage vacuum unit which concentrates the effluent to about 25% total solids. The other uses a four-stage vacuum unit giving a similar result. Again, the final sludges are liquid, and it is questionable whether the same type of evaporators could be used to produce an 80% concentrate suitable for landfill.

Evaporation plants, especially those designed to deliver a very concentrated residue, have unavoidable problems in erosion and fouling of heat-transfer surfaces. Capital costs are high, but running costs may be reasonable if the design includes adequate heat conservation features.

Flocculation with Polyvalent Cations

To reduce costs, many processes have been tried industrially using waste materials such as ferrous sulfate (from steel pickling), aluminum chloride (from petrochemical catalysts), bitterns (the magnesium-rich residue from solar evaporation of seawater), aluminum chlorohydrate (from aluminum refining), and calcium chloride. They have all been found unsatisfactory in several respects, not the least being the relatively high cost for a rather low reduction in pollution. The high cost arises not from the chemical flocculation itself, but from the associated technology of sludge dewatering and disposal. Very few installations of this type have survived, and it is doubtful whether they would be considered today for a new installation.

Hot Acid Flocculation

Two pilot-plant installations of this process have been extensively tested, and it appears to be suitable for mill use [60]. The wastewater is mixed with about 1% of sulfuric acid and passed through a reactor at 100°C, where a breakdown of the structure occurs. The treated liquor is desludged in a decanter centrifuge, giving virtually complete removal of grease and suspended solids. The sludge is reasonably solid, with a solids content around 35%. The liquid is clear but brown in color, and is likely to have a COD of 3000-10000 mg/l. The process is among the cheapest.

Ultrafiltration

One plant has been operating in a scouring mill in the United Kingdom for a few years. The liquors are circulated through membranes which retain the dirt and grease and a small proportion of the dissolved solids but allow most of the water and dissolved solids to pass. The sludge concentrate contains only about 15% of solids and is evaporated to less than 10% water in a rotary evaporator.

The life of the membrane has been shorter than expected, and the effluent still has a high pollution level. The cost appears to be high in relation to the performance, and it is doubtful whether any similar installations will be made. Several workers claim to have made advances in membrane technology which will lead to cheaper and more efficient membrane processes, but these have barely reached the pilot-plant stage.

Solvent Extraction

Treatment of scouring wastewater with a partially water-miscible alcohol such as pentanol, hexanol, or cyclohexanol results in extraction of the grease into the alcohol layer and a ready separation of the suspended solids. Two commercial plants have been installed, one in New Zealand using n-pentanol and one in Czechoslovakia using n-hexanol [67]. Very little information has been published on the performance of these plants, but it is understood that they are technically complex, of high capital cost, and of rather poor performance (most of the suint is discharged in the effluent, giving a COD up to 10,000 mg/liter).

Other Methods

The Lo-flo process. In this process [23], the number of bowls in a scouring machine is increased by substituting a number of small bowls for the first one or two conventional bowls (Fig. 5.15). The water input to the small bowls is reduced to less than 1 liter per kg, thereby increasing the dissolved solids concentration to 10% or more. This causes the liquor to destabilize, and the grease and dirt can then be removed by centrifuging, the liquor being returned to the scouring bowl.

The sludge contains about 50% solids and is disposable by landfill. Grease recovery is roughly doubled compared with conventional scouring. There is no liquid effluent when running with some wool types; with others, a small volume (0.5-1 liter per kg) needs to be disposed of by other means.

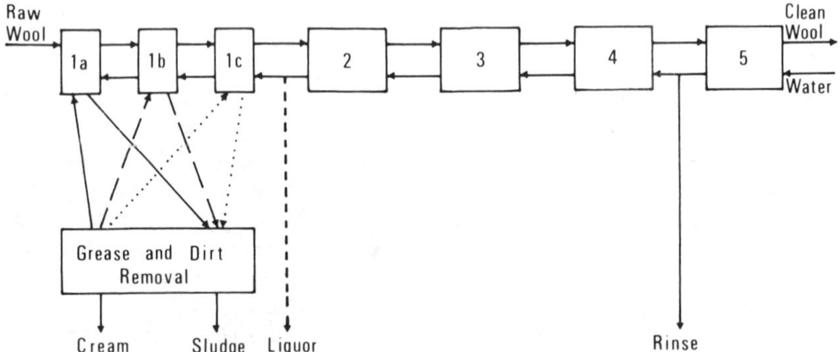

Figure 5.15 The Lo-flo process. Three small bowls, 1a, 1b, and 1c, have been substituted for the first bowl of a five-bowl scouring set, and operate with very high dissolved solids contents. The liquor discharge from bowl 2 may be zero with some wool types. Liquors from all three small bowls are circulated in turn through the centrifuges.

This is potentially the cheapest of all pollution reduction processes. At the time of writing, a mill installation in the United Kingdom was reported to be operating satisfactorily. The latest assessment (mid-1981) is that removal of dirt and suint will probably not exceed 60%, and the main use for the process is likely to be as an adjunct to other treatment processes.

The WRONZ minibowl system. This is an extension of the WRONZ scouring system (Sec. 2.7) in which small scouring bowls of advanced design are used to obtain maximum scouring efficiency. Because of the design of the bowls, the minimum flowdown is about 1.5 liter per kg, which must be evaporated or otherwise treated before discharge. At these water input rates, the liquors are at least partially destabilized and grease recoveries are significantly increased. The sludge from the HST is usually dewatered in a decanter centrifuge. WRONZ minibowls are widely used as ordinary scouring bowls, where they make no contribution to effluent treatment but have other advantages [68].

Treatment of combined wastewaters. In mills where the wool is processed after scouring, effluents are discharged from dyeing, backwashing, and other wet processes. Frequently the combined volume of these effluents amounts to 200-500 liters per kg, and they are quite weak. In these circumstances it is often economical to combine the scouring effluent with the others and treat the whole by conventional methods. The most common method is a combination of biological and chemical processes, followed by sludge dewatering. Many mills round the world have wastewater disposal systems of this sort [69].

The SOVER process. The SOVER solvent scouring process [17] does not produce any polluted discharges, although in ordinary circumstances some waste heat would have to be discharged.

5. ENERGY CONSERVATION

Energy costs (heat, light, and power) make up 15-20% of total scouring costs, a proportion that is likely to rise steeply in the next few years. Energy conservation is therefore likely to become of more importance. Of the different types of energy, power and lighting give little scope for economy, but quite large savings are possible in heat usage. Since heat energy makes up 80-90% of the total energy use, such savings can be significant.

The wool dryer uses about a half or a little more of the total scouring heat energy, and represents the biggest single potential saving. Reductions in dryer heat usage can be made by installing heat exchangers to exchange heat between incoming and exhaust air streams, by operating the final squeezing rollers of the scouring machine at the optimum temperature, and by improving dryer efficiency, that is, using less heat per unit weight of water evaporated.

Another large source of heat loss is in hot liquor discharged to drain. The cost of heat exchangers to recover this heat by exchanging with incoming cold water is usually saved in a year. A reduction in water consumption will automatically reduce the quantity of heat discharged with the wastewater.

If the simplest of the above measures are combined with good housekeeping (avoiding steam leaks, etc.), a saving in heat energy of around 30% is possible, giving an overall cost reduction on the order of 4-5%. (See Barker [48] for a recent discussion of this subject.)

6. SOLVENT SCOURING

6.1 General Considerations

Agitating wool fiber assemblies in anhydrous solvents does not cause felting. Many comparative tests have shown that in topmaking the quantity of noil produced when the wool has been solvent scoured is only three-quarters of that produced after normal aqueous scouring. The yield of scoured wool from solvent scouring is nearly always 0.5-1.0% higher than from aqueous scouring. These advantages add up to several tens of thousands of dollars per year for a mill of moderate size, and are the basis for a long continuing interest in solvent scouring.

A principal difficulty in solvent scouring is that solvents that dissolve the grease do not remove water-soluble contaminants. Furthermore , the detergency mechanisms that remove dirt particles in aqueous scouring do not operate when a solvent is used. Consequently, it is extremely difficult to clean wool to commercial standards using solvents. In practice, these difficulties have been tackled in two ways: either two or more solvents (including water) have been used, or the wool has been intensely agitated. Both these features have been present in most commercial solvent scours. Where two solvents are used, one is a nonpolar solvent for removing wool grease and the other is a polar solvent for removing suint.

Another serious problem is in separating aqueous dirt from solutions of wool grease in nonpolar solvents. Raw wool always has a moisture content of about 10% (Table 5.1), and part of this water can form complex emulsions with the wool grease solution. These emulsions must be broken before the solvent is recovered by distillation, but the only practical way of doing this is to add a water-miscible solvent, usually one of the lower alcohols. This naturally leads to a three-component extraction using a nonpolar solvent, an alcohol, and water. Handling of systems of this sort is very complex. To keep solvent losses low and achieve the required separations of impurities and solvents requires an elaborate and expensive plant.

The development of solvent scouring is illustrated by the descriptions of various processes that follow. All simple processes have eventually been discontinued, and the only process currently operating is the SOVER process [17].

6.2 Discontinued Solvent Scouring Processes

Extraction in Packed Vessels

This process was operated for a number of years by Arlington Mills in the United States and Solvent Belge in Belgium [4]. The wool was packed into large vessels and extracted with several washes of hexane or a similar solvent. The close packing of the wool prevented an appreciable quantity of dirt from entering the solvent. The grease content of the wool was reduced to about 1%. After the residual solvent had been removed from the wool by blowing hot air through it, the wool was scoured through a conventional scouring train. The grease was recovered from the solution by distilling off the solvent.

This process is degreasing rather than scouring. Its advantages were that grease recovery was very high, costs were reasonable because little if any detergent was required in the conventional scour, and the wool was less felted owing to the very mild action required in the conventional scour. Because this was a batch process, labor costs were high and it became progressively less suitable for modern conditions.

Solvent Jet Scouring

The CSIRO process of the early 1950s [70] was operated for more than a decade by Michell in South Australia. The wool was carried on long, woven wire conveyors underneath a number of jets of white spirit, then under water jets.

The idea behind the process was that the energy of the jets would displace all the insoluble contaminants from the wool, and the water jets would displace the residual solvent and dissolve any suint not displaced by the solvent. Removal of the solvent from the wool before it was dried overcame the hazard of drying flammable solvents. Since white spirit and water are completely immiscible, separation of the two phases should have been simple.

In practice, the cleaning action was found to be inadequate, and additional cleaning was required in backwashing after carding. The process could not be used for nonbackwashed wools. Carryover of abrasive particles into carding and combing was also a problem. The solvent and water formed intractable emulsions, and solvent losses to both drain and the atmosphere were high.

Nevertheless, the solvent was cheap, grease recovery was high, and there were substantial advantages in combing tear improvement. The process could not be used in today's climate of high petroleum prices and strict environmental controls. The Michell plant was probably the first commercial scouring plant to use a decanter centrifuge for dirt removal—in this case from solvent.

The Alfa Laval Process

This process [71] used kerosene as a solvent and relied on good opening of the wool and a high level of agitation to remove the impurities. Water was introduced during solvent recovery to remove insoluble and water-soluble contaminants from the solvent. Flammability problems were minimized by using a solvent with high flash point.

A plant was operated for a fairly short time at Whitehead's mill in the United Kingdom. The continuous process involved a combination of mixing, jetting, and vibrating conveyers to clean the wool. The problems were the usual ones of poor wool cleaniness, difficult emulsion separations, and solvent losses.

Dry-Cleaning Procsses

Two processes using dry-cleaning machinery for batchwise scouring of greasy wool made brief commercial or semicommercial appearances during the 1960s. In one, operated by Brown-Gouge in Australia [72], batches of greasy wool were loaded into open-weave bags and dry-cleaned with the usual dry-cleaning solvents—trichlorethylene or perchlorethylene containing a small quantity of water and detergent. The degreased wool was then washed through two or three scouring bowls containing water.

The other process, at Burnley's mill in the United Kingdom [73], was developed to a greater degree of sophistication. Tetrachlorethylene was the solvent, and isopropanol was used to facilitate solvent-water separation.

Being batch processes, neither of these fitted well into modern production requirements.

6.3 The SOVER Process

This process [17] must surely be the last word in solvent scouring! It is a continuous jetting process in which all of the traditional solvent scouring problems of flammability, emulsion separation, solvent purification, and solvent conservation have been solved by advanced technology. The wool is carried on a continuous steel belt conveyor through jets of water, isopropanol, and hexane, in that order. Jetting by the two organic solvents is extensive and the wool is claimed to be adequately cleaned. The liquids are combined in proportions that lead to a gravity separation into three layers. The top layer is a solution of wool grease in hexane, the bottom layer is a solution of suint in water-isopropanol, and the middle layer contains the insoluble impurities. (Actually all three layers contain all three components, the solvents indicated being the major component.)

The three layers are worked up in a fairly complicated fashion, the grease being neutralized and bleached while still in solution. The final products are high-grade wool grease, wool grease fatty acids, coarse dirt, and a fertilizer comprising the fine dirt, suint, and phosphoric acid derivatives remaining after the acid splitting of soaps formed during neutralization.

To be economic, the SOVER process must be run continously at a high throughput. The economics also depend rather heavily on the sale of wool grease—a notoriously fickle market. An advantage today over aqueous scouring processes is that the SOVER process does not produce any polluting discharges.

The very high cost (approaching $10 million in 1980) has been the major factor limiting the sales of the process. At mid-1980 only three plants had been sold, one to Japan and two to the USSR.

7. OTHER METHODS OF CLEANING RAW WOOL

For the sake of completeness, brief mention is made of some ideas that have been promoted at various times. Only the first enjoyed a brief commercial run, and utilization of any of them in the future must be considered extremely doubtful.

7.1 Chilling and Dusting

When raw wool is cooled, the wool grease becomes brittle at a temperature where the wool itself is still flexible. If the chilled wool is dusted, much of the grease, suint, and associated impurities can be separated from it as dust. This process was operated for a time by the Frosted Wool Co. in the United States [74], primarily as a means of removing vegetable matter. The degree of cleanliness obtained was not good, and the wool required a conventional scouring process before carding.

7.2 Treatment with Powders

The original fulling process for cleaning wool consisted of rubbing a dry powdered clay (*fuller's earth*) into the wool. Grease and suint were sorbed by the clay, which was then shaken out of the wool.

A bizarre process using this principle was proposed fairly recently by Boer [75]. A cereal powder is used in a countercurrent process operating through the opening, carding, and gilling stages of wool processing. Clean powder is mixed with the wool during the third gilling before combing, and separated from it afterward. This powder is mixed with the wool during the second gilling process, and again separated. In this way the powder works its way backward through the gilling and carding operations, its first contact with raw wool being in the opening machines.

As the clean powder works its way backward through the processing it picks up more and more of the impurities from the wool. The wool, of course, becomes progressively cleaner, until it is commercially clean after the third gilling. The fascinating novelty in the process is that the contaminated cereal is fed to sheep! The only products from this brilliant natural cycle are meat, wool, and manure, the perfect answer to the waste disposal problem! A scientific study of extraction with sorbent powders was published [76] in the 1970s.

7.3 Processing Wool Without Scouring

Attempts have been made from time to time to form raw-wool slivers, which could then be cleaned by a process similar to backwashing [28, 77]. The rationale behind such a procedure is that it is potentially more efficient to form slivers from raw wool in which there is already a high degree of fiber parallelism, rather than to deparallelize the fibers in scouring and then reparallelize them in carding. The many practical difficulties in such a process have not been satisfactorily overcome.

REFERENCES

1. H. B. Carter and W. H. Clarke, *Aust. J. Agric. Res. 8,* 91, 109 (1957).
2. C. A. Anderson and G. F. Wood, *Nature 193,* 742 (1962).
3. C. A. Anderson and G. F. Wood, *Proceedings of the 3rd Int. Wool Text Res. Conf.*, Vol. 3, p. 129 (1965).
4. H. V. Truter, *Wool Wax,* Cleaver-Hume Press, London (1956).
5. K. Motiuk, *J. Amer. Oil Chem. Soc. 56,* 91 (1979).
6. K. Motiuk, *J. Amer. Oil Chem. Soc. 56,* 651 (1979).
7. K. Motiuk, *J. Amer. Oil Chem. Soc. 57,* 145 (1980).
8. R. G. Stewart and L. F. Story, Technical Papers No. 4, WRONZ, Christchurch, New Zealand (1980).
9. C. A. Anderson and J. R. Christoe, submitted to *J. Text. Inst.*
10. T. E. Mozes and D. W. F. Turpie, *SAWTRI bull. 13* (2), 43 (1979).
11. H. Zahn, J. Fohles, and M. Niehaus, *Proceedings of the F.L.I. London meeting, June 1979,* Report No. 16, Federation Lainere International, Paris (1979).

12. M. Lipson and U. A. F. Black, *Proc. Roy. Soc. N.S.W. 78,* 84 (1944).
13. J. R. McCracken and M. Chaikin, *J. Text. Inst. 62,* 633 (1971).
14. G. F. Wood, *J. Text. Inst. 60,* 249 (1969).
15. S. Neubart, *Industrie Text.* (Jan.), 15; (Feb.), 77 (1965).
16. G. V. Barker and R. G. Stewart, WRONZ Communication No. 61 (1978).
17. J. Brach, *Wool Sci. Rev.* No. 36, 38 (1969).
18. A. Samson and M. Chaikin, *Proceedings of the 3rd Int. Wool Text. Res. Conf.*, Vol. 3, p. 151 (1965).
19. C. A. Anderson, M. Lipson, J. F. Sinclair, and F. G. Wood, *Proceedings of the 3rd Int. Wool Text. Res. Conf.*, Vol. 3 p.141 (1965).
20. *Text J. Australia, 44,* June, 34 (1969).
21. R. Bownass, WIRA Report No. 47, Leeds, England (1969).
22. *Text. Recorder, 81* (970), 69 (1964).
23. G. F. Wood, A. J. C. Pearson, and J. R. Christoe, Report No. G39, CSIRO Division of Textile Industry, Geelong, Australia (1979).
24. R. G. Stewart, *J. Text. Inst. 54,* T88 (1963).
25. J. R. McCracken, *Inst. Engrs. Australia, Mech. and Chem. Eng. Trans.* 1973, 25.
26. G. V. Barker, J. Murrow, and R. G. Stewart, WRONZ Communication No. 15 (1973).
27. J. V. Poncelet, *Text. J. Australia 39* (Sept.), 30 (1964).
28. M. Chaikin and J. R. McCracken, *Interwool 71,* Wool Research Institute, Brno, Czechosovakia, p. 8 (1971).
29. Y. Matsuzaki and J. Sekiguchi, *J. Text. Machinery Soc. Jap., 19* (4-5), 110 (1973).
30. M. S. Nossar, *Proceedings, DISC 80,* Australian Wool Corporation, Melbourne p. 57 (1980).
31. E. C. Hansen, *Amer. Dyestuff Reptr. 47,* 155 (1958).
32. E. J. McNamara, *Amer. Dyestuff Reptr. 46,* 731 (1957).
33. R. A. Olney and B. A. Ryberg, *Amer. Dyestuff Reptr. 45,* 781 (1956).
34. H. C. Borghetty, *Amer. Dyestuff Reptr. 44,* 726 (1955).
35. H. J. Henning, *Z. Gesamte Text. Ind. 69,* 237 (1967).
36. C. A. Anderson and G. F. Wood, *J. Text. Inst. 57,* T545 (1966).
37. G. F. Wood, *Proceedings, DISC 80,* Australian Wool Corporation, Melbourne (1980).
38. G. Jones, *J. Appl. Chem. Biotechnol. 21,* 39, 48 (1971).
39. R. L. Elms and G. F. Wood, *Text. Inst. Industry 17,* 210 (1979).
40. J. L. Hoare, *J. Text. Inst. 65,* 500 (1974).
41. S. D. Rossouw, *S. Afr. Indust. Chemist* No. 6, 90 (1952).
42. R. G. Jamieson, *Text. Inst. Industry 17,* 70 (1979).
43. C. A. Anderson, *J. Text. Inst. 53,* T401 (1962).
44. D. B. Early, *J. Text. Inst. 70,* 518 (1979).

45. G. V. Barker, WRONZ Communication No. 21 (1973).
46. R. G. Stewart, G. V. Barker, P. Chisnall, and J. L. Hoare, WRONZ Report No. 25 (1974).
47. *WIRA Bulletin 13,* 38 (1950).
48. G. V. Barker, *Proceedings DISC 80,* Australian Wool Corporation, Melbourne (1980).
49. B. H. Mackay, *J. Text Inst. 54,* T376 (1963).
50. J. L. Hoare and B. Thompson, *J. Text. Inst. 65,* 281 (1974).
51. G. F. Wood, *Text. Inst. Industry 14,* 113 (1976).
52. C. A. Anderson, *Text. Res. J. 30,* 51 (1960).
53. L. G. Vasileva, *Nauch. Issled. Tr. Tsent. Nauch. Issled. Inst. Sherst. Prom 24,* 26 (1970).
54. C. A. Anderson and G. F. Wood, *J. Text Inst. 57,* T55 (1966).
55. F. V. Wells and I. I. Lubowe, *Soap Perfum. Cosmet. 34,* 637 (1961).
56. H. Wagner, *Amer. Perfumer Aromat. 75,* 23 (1960).
57. M. L. Schlossman and J. P. McCarthy, *J. Amer. Oil Chem. Soc. 55,* 447 (1978).
58. C. A. Anderson and G. F. Wood, *J. Amer. Oil Chem. Soc. 40,* 333 (1963).
59. R. G. Jamieson and R. G. Stewart, WRONZ Communication No. 53, (1977).
60. J. R. McCracken, *Prog. in Water Technol. 10,* 503 (1978).
61. E. Hurwitz, G. H. Teletzke, and W. B. Gitchel, *Water and Sewage Works 112,* 298 (1965).
62. A. J. C. Pearson, C. A. Anderson, and G. F. Wood, *J. Water Poll. Control Fed. 48,* 945 (1976).
63. N. C. Beaton, *Text. Inst. Industry 13,* 361 (1975).
64. T. E. Mozes and D. W. F. Turpie, SAWTRI Tech. Report No. 349 (1977).
65. T. E. Mozes and D. W. F. Turpie, SAWTRI Tech. Report No. 386 (1978).
66. *Japan Wool Textile Industry's Treatment of Effluent,* Japan Wool Spinners' Association, Tokyo (1973).
67. J. Valenta and M. Koubik, *Interwool 71,* Wool Research Institute, Brno. Czechoslovakia, p. 171 (1971).
68. P. E. Chisnall and R. G. Stewart, *Text. Inst. Industry 17,* 68 (1979).
69. *Rulemaking for the Textile Mills. Point Source Category,* Effluent Guidelines Division, E.P.A., Washington, D.C. (1979).
70. Report No. G10, CSIRO Division of Textile Industry, Geelong, Australia (1960).
71. Technical Bulletin No. 5801E, Alfa-Laval Co., Stockholm (1960).
72. *Wool Record 107* (2927), 28 (1965).
73. N. Saville, W. J. Shelton, R. Ward, and J. Sewell, *Appl. Polym. Symp. 18,* 1157 (1971).

74. R. J. Wig, *Amer. Dyestuff Reptr. 24,* 270 (1935).
75. A. Boer, U.S. Patent 2,982,676 (1961).
76. G. H. Robertson and J. P. Morgan, *Ind. Eng. Chem. Process Des. Devel. 14,* 12 (1975).
77. D. Walker, British Patent 1,287,741 (1972).

INDEX

Active surface of fibers, 57
Adsorption at fiber-liquid interface, 63-64
Adsorption at solid-liquid interface, 59-63
Alkali hydroxide ions, solvated forms of, 95
Alkali treatment of cellulose fibers, 93-165
  action of alkali agents on cellulose fibers, 94-111
  mercerization of cotton fibers, 134-157
  scouring of cotton, 111-134
Anhydrous liquid ammonia, physical constants of, 168
Anionic detergents, 221-222
Aqueous systems, interaction with fiber of, 51-92
  general considerations, 52-55
  mass transfer in textile chemical processes, 70-72
  penetration, 72-91
  sorption processes in textile chemistry, 55-70

Blend fabrics, mercerization of, 149-152

Case II diffusion, 14-15
Cellulose:
  mercerized, effect of liquid ammonia on, 175-176
  thermokinetics of liquid penetration into, 85-89
Cellulosics, liquid ammonia treatment of, 176-190
Centrifuging, 236-239
Color-woven goods, scouring of, 133-134
Cotton:
  effect of liquid ammonia on structure and morphology of, 170-175
  mercerization of, 134-157
  scouring of, 111-134
Cotton ash, composition of, 119

Detergent scouring, 213-233
Detergent systems, types of, 219-223
Detergents, utilization of, 222-223
Differential scanning calorimetry (DSC), 32
Dimethylformamide (DMF), 23, 25, 29, 30, 31, 42

N,N'-Dimethylol-4,5-dihydroxy-ethyleneurea (DMDHEU), 190, 191, 192, 193
N,N'-Dimethylolethyleneurea (DMEU), 190
Dyeing:
  from liquid ammonia, 193-195
  solvents in, 34-44
Dynamic shrinkage, 22-26

Energy conservation, 249-250

Fabrics:
  blend, mercerication of, 149-152
  knitted, mercerization of, 157
  liquid ammonia treatment of, 180-190
  woven, mercerization of, 153-155
Fatty acids found in cotton wax, 116
Fatty alcohols found in cotton waxes, 117
Felting in scouring, 210-211
Fiber penetration, 72-74
Fibers:
  active surface area of, 57
  noncellulosic, liquid ammonia treatment of, 195-198
  polymeric, diffusion of organic solvents into, 9-15
Fickian diffusion, 10-12
Finishing operations, solvents in, 34-44
Flame-resist finishing, 192-193
Fleissner scouring bowl, 216
Flory-Huggins theory, 6-7

Heat of immersion of fibers in water, 57
Hot mercerization process, 146-149
Hydronamic scour, 218-219

Industrial scouring, wastewater disposal, 246-249
Isothermal shrinkage, 20-22

Jet scouring bowl, 217
Jig process, 127-132

Knitted fabrics, mercerization of, 157

Liquid ammonia, anhydrous, physical constants of, 168
Liquid ammonia treatment of textiles, 167-203
  dyeing from liquid ammonia, 193-195
  effect of liquid ammonia on rayon or mercerized cotton, 175-176
  effect of liquid ammonia on structure and morphology of cotton, 170-175
  equipment for, 198
  future of, 198-199
  liquid ammonia in the application of topical finishes, 190-193
  liquid ammonia treatment of cellulosics, 176-190
  liquid ammonia treatment of fabrics, 180-190
  liquid ammonia treatment of noncellulosic fibers, 195-198
  properties of liquid ammonia, 168-169

Liquor systems, 225-231
Lo-flo process, 218

Mass transfer in textile chemical processes, 70-72
Mercerization of cotton fibers, 134-157
Mercerization of yarns, 152-153
Mercerized cellulose, effect of liquid ammonia on, 175-176

Noncellulosic fibers, liquid ammonia treatment of, 195-198
Non-Fickian diffusion, 12-14
Nonionic detergents, 220-221
Nylons, liquid ammonia treatment of, 198

Organic solvents, interactions between fibers and, 1-49
  diffusion of organic solvents into polymeric fibers, 9-15
  solvent induced changes in the physical properties of fibers, 27-34
  solvent induced modification of fiber structure, 15-19
  solvent induced shrinkage in oriented polymeric systems, 20-27
  solvents in dyeing and finishing operations, 34-44
  thermodynamic considerations in fiber-solvent interactions, 3-9

Plasticization of polymeric fibers by solvents, 28-30
Polymeric fibers, diffusion of organic solvents into, 9-15
Potassium hydroxide hydrates, 96
Prewetting followed by penetration of fabrics, 76-84

Raw wool:
  other methods of cleaning, 253-254
  properties of, 206-209
  scouring, 205-213
Rayon, effect of liquid ammonia on, 175-176

Scouring of color-woven goods, 133-134
Scouring of cotton, 111-134
Scouring machines, traditional, 213-216
Scouring, wastewater disposal, 241-249
Shrinkage:
  dynamic, 22-26
  isothermal, 20-22
Soap, 219-220
Soda celluloses, 99
Sodium hydroxide, concentration and temperature of, maximum solubility as a function of, 105
Sodium hydroxide concentration, maximum swelling as a function of, 102
Sodium hydroxide hydrates, 95
Solvent induced crystallization (SINC), 16-17
Solvent scouring, 250-253
Sorption processes in textile chemistry, 55-70
SOVER process, 252-253
Specific free surface energy (SFSE), 55
  of fibers, 56

Suint scouring, 223
Swelling processes, comparison of, 184-185

Thermodynamic considerations in fiber-solvent interactions, 3-9
Thermokinetics of liquid penetration into cellulose, 85-89
Topical finishes, liquid ammonia in the application of, 190-195

Wool grease recovery, 233-241
Wool grease refining, 239-241
Wool scouring streams, distribution of components in, 242
Wool:
  liquid ammonia treatment of, 195-197
  illustrative compositions of, 209
Woven fabrics, mercerization of, 153-155

Yarn:
  liquid ammonia treatment of, 177-180
  mercerization of, 152-153